Residential
Heat Pumps

RESIDENTIAL HEAT PUMPS: Installation and Troubleshooting

S.E. Sutphin

PRENTICE-HALL, INC.
Englewood Cliffs, New Jersey 07632

Library of Congress Cataloging-in-Publication Data

Sutphin, S. E. (date)
   Residential heat pumps.

   Includes index.
   1. Heat pumps.   2. Dwellings—Heating and
ventilation.   3. Dwellings—Air conditioning.
I. Title.
TH7638.S87 1987   697   86-12178
ISBN 0-13-774613-X

Editorial/production supervision and
   interior design: *Carol L. Atkins*
Cover design: *Wanda Lubelska*

Printed in the United States of America

10 9 8 7 6 5 4 3 2

$4/q4 \; 6IF-$

ISBN   0-13-774613-X   025

Prentice-Hall International (UK) Limited, *London*
Prentice-Hall of Australia Pty. Limited, *Sydney*
Prentice-Hall Canada Inc., *Toronto*
Prentice-Hall Hispanoamericana, S.A., *Mexico*
Prentice-Hall of India Private Limited, *New Delhi*
Prentice-Hall of Japan, Inc., *Tokyo*
Prentice-Hall of Southeast Asia Pte. Ltd., *Singapore*
Editora Prentice-Hall do Brasil, Ltda., *Rio de Janeiro*

The author would like to gratefully acknowledge both Copeland Corporation and Carrier Corporation, who provided a vast amount of information for research purposes during the writing of this book. Without their valuable assistance, this work would not have been possible.

Contents

          WATER-SOURCE HEAT PUMPS   **116**

          5.1 Design Considerations: Ground-water
               Heat Pump   *117*

               Type and Condition of an Existing Well   117
               Water Quality   119
               Return Well Considerations   121
               New Supply and Return Wells   121

          5.2 Design Considerations: Closed-loop
               Heat Pumps   *122*

          5.3 Installing the Heat Pump   *124*

               Location   124
               Ductwork   125
               Mounting the Heat Pump   126
               Electrical Connections   126
               Location of the Thermostat   127
               Water Piping Accessories and Considerations   127
               Condensate Drain Piping   131
               Flowmeters   131
               Pete's Plugs   131

          5.4 Closed-loop Pump Kits   *132*

               Flushing and Testing a Closed Loop   134
               *Charging a Closed-loop System with
               Antifreeze   136*
               *Checking the Flow Rate in a Closed-loop
               System   137*

          5.5 Start-up   *138*
               Checklist   138
               Start-up Procedure   139

          5.6 Installation: Water-to-water Units   *140*

SIX   INSTALLATION AND START-UP:
          AIR-SOURCE HEAT PUMPS   **142**

          6.1 Location and Mounting of the Outdoor
               Unit   *143*

               Ground-level Installation   143
               Rooftop Installation   143

          6.2 Location of the Indoor Unit   *144*
               Oil Return Considerations   145

Preface

The heat pump, although not a new development, has become increasingly popular in the last decade, particularly since the Arab oil embargo of the seventies which forced many to consider alternatives to the use of oil for heating purposes. The second major factor that has contributed to the increase in popularity of heat pumps is an all-out campaign by power companies to promote their use. This was a necessary move on the part of power companies, since they normally experience much higher power loads during the summer months due to increased use of air conditioning units; and it was and is their intention to attempt to equalize power loads by their promotion of the dual-purpose heat pump as an alternative to two separate systems, one for heating and another for cooling.

The heat pump is often referred to as a "reverse-cycle air conditioner." This is because it has the unique ability to reverse its mode from heating to cooling and vice versa on demand. This ability is directly related to the same principles that have been utilized in air conditioning systems for many years, principles that are based upon the laws of thermodynamics, or the mechanical action of heat. Through many years of research, heat pump manufacturers have developed an efficient, yet relatively inexpensive system, that can be installed as a stand-alone unit or as an add-on to an existing heating system.

Although there are two major types of heat pumps—air-source and water-source, the major push for their use concentrated primarily on the air-source units until very recently. However, the air-source heat pump, which utilizes the outside air for its heat source, almost always requires a back-up heating system to compensate for periods when the unit must work extremely hard to obtain a comparatively small amount of heat when outside temperatures are very cold. Because of this, many who might otherwise have purchased a heat pump became disillusioned with this so-called efficient machine and looked elsewhere for a more conventional means of heating and/or cooling their homes.

Enter the water-source heat pump. Again not a new development, but one that had been erroneously moved to the back burner because of its dependence on an adequate underground water supply, something that many felt just was not available. This theory has since been dispelled and new research and development are now bringing the water-source heat pump to the forefront. All research indicates that the water-source heat pump, which utilizes a year-round, constant-temperature water source, is more efficient than its air source counter-part and often does not require a back-up heating system.

The type of heat pump chosen for a particular residence, however, depends on a great many factors; and it is the responsibility of the installer to carefully weigh all factors and assist the homeowner in making the proper decisions with regard to the installation of a heat pump. This book is intended to provide the reader with a thorough background into the principles utilized in the modern-day air-source and water-source heat pump. Also included are chapters specifically relating to the pros and cons of installation, maintenance, and troubleshooting of both types. It is hoped that this book, which takes a long, hard look at very recent and very positive innovations with regard to the water-source heat pump in particular, will provide the technician with the information necessary for him to:

1. Understand the principles utilized in both air-source and water-source heat pumps, including the sub-categories in each of these major categories.
2. Determine the appropriate system for a particular residence.
3. Install the system properly.
4. Aid the homeowner in maintaining the system.
5. Troubleshoot and repair the system, component by component.

S. E. Sutphin

One
Fundamentals of Refrigeration

Any discussion of heat pumps must begin with a study of the fundamentals of refrigeration, since heat pumps are derived from standard air conditioning and refrigeration equipment. Although most persons normally associate refrigeration with cold and cooling, the practice of refrigeration engineering deals almost entirely with the transfer of heat. This apparent paradox is one of the fundamental concepts that must be grasped in order to understand the workings of a refrigeration system. These same concepts can then be transferred to the workings of a heat pump.

1.1 THERMODYNAMICS

Thermodynamics is that branch of science dealing with the mechanical action of heat. These fundamental principles of nature are basic to the study of refrigeration.

The first and most important of these laws is the fact that *energy can neither be created or destroyed*, but can be converted from one type of energy to another. A study of thermodynamic theory is beyond the scope of this book, but the examples that follow will illustrate the practical applications of this energy law.

Heat

Heat is a form of energy, primarily created by the transformation of other types of energy into heat energy. For example, mechanical energy turning a wheel causes friction, which creates heat.

Heat is often defined as *energy in transfer*, because it never stands still, but is always moving from a warm body to a colder body. Much of the heat on the Earth is derived from radiation from the sun. A spoon in ice water loses its heat to the water and becomes cold; a spoon in hot coffee absorbs heat from the coffee and becomes warm. But the terms "warmer" and "colder" are only comparative. Heat exists at any temperature above absolute zero, even though it may be in extremely small quantities. *Absolute zero* is the term used by scientists to describe the lowest theoretical temperature possible, the temperature at which no heat exists, which is approximately $460°$ below Fahrenheit. By comparison with this standard, the coldest weather we might ever experience on Earth is much warmer.

Temperature

Temperature is the scale used to measure the intensity of heat, the indicator that determines which way the heat energy will move. In the United States, temperature is normally measured in degrees Fahrenheit, (F) but the Centigrade (C) scale (sometimes termed Celsius) is widely used in other parts of the world. Both scales have two basic points in common, the freezing point of water, and the boiling point of water at sea level (see Figure 1.1). Water freezes at $32°F$ and $0°C$, and water boils at sea level at $212°F$ and $100°C$. On the Fahrenheit scale, the temperature difference between these two points is divided into 180 equal increments or degrees F, while on the Centigrade scale, the temperature difference is divided into 100 equal increments or degrees C. The relation between Fahrenheit and Centigrade scales can be established by the following formulas:

$$\text{Fahrenheit} = 9/5 \text{ Centigrade} + 32°$$

$$\text{Centigrade} = 5/9 \text{ (Fahrenheit} - 32°)$$

Heat Measurement

The measurement of temperature has no relation to the quantity of heat. A match flame may have the same temperature as a bonfire, but obviously the quantity of heat given off is vastly different.

The basic unit of heat measurement used today in the United States is the British Thermal Unit, commonly expressed as a BTU. A

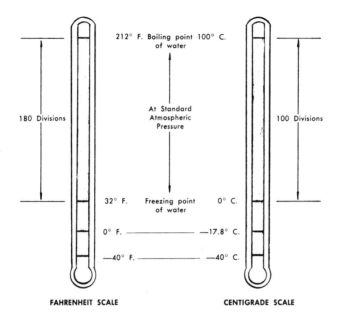

Figure 1.1   Temperature scales (Courtesy Copeland Corporation).

BTU is defined as the amount of heat necessary to raise 1 pound of water 1 degree Fahrenheit. For example, to raise the temperature of 1 gallon of water (approximately 8.3 pounds) from 70°F to 80°F will require 83 BTUs.

1.2 HEAT TRANSFER

The second important law of thermodynamics is that *heat always travels from a warm object to a colder one.* The rate of heat travel is in direct proportion to the temperature difference between the two bodies.

Assume that two steel balls are side by side in a perfectly insulated box. One ball weighs 1 pound and has a temperature of 400°F, while the second ball weighs 1,000 pounds and has a temperature of 390°F. The heat content of the larger ball is tremendously greater than the small one, but because of the temperature difference, heat will travel from the small ball to the large one until the temperatures equalize.

Heat can travel in any of three ways: radiation, conduction, or convection.

Radiation *is the transfer of heat by waves* similar to light waves or radio waves. For example, the sun's energy is transferred to the Earth by radiation. One need only step from the shade into direct sunlight to feel the impact of the heat waves, even though the temperature of the surrounding air is identical in both places. There is little radiation at low temperatures and at small temperature differences, so radiation is of little importance in the actual refrigeration process. However, radiation to the refrigerated space or product from the outside environment, particularly the sun, may be a major factor in the refrigeration load.

Conduction *is the flow of heat through a substance.* Actual physical contact is required for heat transfer to take place between two bodies by this means. Conduction is a highly efficient means of heat transfer, as anyone who has touched a piece of hot metal can relate.

Convection *is the flow of heat by means of a fluid medium*, either gas or liquid, normally air or water. Air may be heated by a furnace and then discharged into a room to heat objects by convection.

In a typical refrigeration application, heat normally will travel by a combination of processes; the ability of a piece of equipment to transfer heat is referred to as *the overall rate of heat transfer.* While heat transfer cannot take place without a temperature difference, different materials vary in their ability to conduct heat. Metal is a very good heat conductor, while asbestos has so much resistance to heat flow that it can be used as insulation.

1.3 CHANGE OF STATE

Most common substances can exist as a solid, a liquid, or a vapor (gas), depending on their temperature and the pressure to which they are exposed. Heat can change their temperature and also can change their state. Heat is absorbed even though no temperature change takes place when a solid changes to a liquid, or when a liquid changes to a vapor. The same amount of heat is given off when the vapor changes back to a liquid, and when the liquid is changed to a solid.

A common example of this process is water, which exists as a liquid, but can also exist in solid form as ice, and as a gas when it becomes steam. As ice, it is a usable form of refrigeration, absorbing heat as it melts at a constant temperature of $32°F$. If placed on a hot stove in an open pan, water's temperature will rise to the boiling point ($212°F$ at sea level). Regardless of the amount of heat applied, its temperature cannot be raised above $212°F$, because the water will completely vaporize into steam. If this steam could be enclosed in a container and more heat applied, then the temperature could again be raised. Obviously, the boiling or evaporating process absorbed heat.

When steam is condensed, it changes back into water, giving off exactly the same amount of heat that it absorbed in the evaporation process. (The steam radiator is a common application of this source of heat.) If the water is to be frozen into ice, the same amount of heat that is absorbed in melting must be extracted by some refrigeration process to cause the freezing action.

The question arises: Where did the heat units go? Scientists have found that all matter is made up of molecules, infinitesimally small building blocks that are arranged in certain patterns to form different substances. In a solid or liquid, the molecules are very close together. In a vapor, the molecules are much farther apart and move about much more freely. The heat energy that was absorbed by the water became molecular energy, and as a result the molecules rearranged themselves, changing the ice into water and the water into steam. When the steam condenses, that same molecular energy is again converted into heat energy.

Sensible Heat

Sensible heat is defined as the heat involved in a change of temperature of a substance. When the temperature of water is raised from 32°F to 212°F, an increase in sensible heat content is taking place. The BTUs required to raise the temperature of 1 pound of a substance 1°F is termed its *specific heat.*

By definition, the specific heat of water is 1.0, but the amount of heat required to raise the temperature of different substances through a given temperature range will vary. It requires only .64 BTU to raise the temperature of 1 pound of butter 1°F, and only .22 BTU is required to raise the temperature of 1 pound of aluminum 1°F. Therefore, the specific heats of these two substances are .64 and .22, respectively.

Latent Heat of Fusion

A change of substance from a solid to a liquid, or from a liquid to a solid involves the *latent heat of fusion.* It might also be termed the latent heat of melting, or the latent heat of freezing.

When 1 pound of ice melts, it absorbs 144 BTUs at a constant temperature of 32°F; if 1 pound of water is to be frozen into ice, 144 BTUs must be removed from the water at a constant temperature of 32°F. In the freezing of food products, it is only the water content for which the latent heat of freezing must be taken into account. This is normally calculated by determining the percentage of water content in a given product.

Latent Heat of Evaporation

A change of a substance from a liquid to a vapor, or from a vapor back to a liquid involves the *latent heat of evaporation.* Since boiling is only a rapid evaporation process, it might also be called the latent heat of boiling, the latent heat of vaporization, or for the reverse process, the latent heat of condensation. When 1 pound of water boils or evaporates, it absorbs 970 BTUs at a constant temperature of $212°F$ (at sea level); to condense 1 pound of steam to water, 970 BTUs must be extracted from it. Because of the large amount of latent heat involved in evaporation and condensation, heat transfer can be very efficient during this process. The same changes of state affecting water apply to any liquid, although at different temperatures and pressures.

The absorption of heat by changing a liquid to vapor, and the discharge of that heat by condensing the vapor, is the keystone to the whole mechanical refrigeration process; the movement of the latent heat involved is the basic means of refrigeration.

Latent Heat of Sublimation

A change in state directly from a solid to a vapor, without passing through the liquid phase, can occur in some substances. The most common example is the use of dry ice or solid carbon dioxide for cooling. The same process can occur with ice below the freezing point, and is also utilized in some freeze-drying processes at extremely low temperatures and high vacuums. The *latent heat of sublimation* is equal to the sum of the latent heat of fusion and the latent heat of evaporation.

Saturation Temperature

The condition of temperature and pressure at which both liquid and vapor can exist simultaneously is termed *saturation.* A saturated liquid or vapor is one at its boiling point; for water at sea level, the saturation temperature is $212°F$. At higher pressures, the saturation temperature increases; and with a decrease in pressure, the saturation temperature decreases.

Superheated Vapor

After a liquid has changed to a vapor, any further heat added to the vapor raises its temperature as long as the pressure to which it is exposed remains constant. Since a temperature rise results, this is sensible heat. The term "superheated vapor" is used to describe a gas whose temperature is above its boiling or saturation point. The air around us is composed of superheated vapor.

Subcooled Liquid

Any liquid that has a temperature lower than the saturation temperature corresponding to its pressure is said to be *subcooled*. Water at any temperature less than its boiling temperature (212°F at sea level) is subcooled.

1.4 PRESSURE

Atmospheric Pressure

The atmosphere surrounding the Earth is composed of gases, primarily oxygen and nitrogen, extending many miles above the Earth's surface. The weight of that atmosphere pressing down on the Earth creates the atmospheric pressure in which we live.

At a given point, the atmospheric pressure is relatively constant, except for minor changes due to changing weather conditions. For purposes of standardization and as a basic reference for comparison, the atmospheric pressure at sea level has been universally accepted, and this has been established at 14.7 pounds per square inch, which is equivalent to the pressure exerted by a column of mercury 29.92 inches high.

At altitudes above sea level, the depth of the atmospheric blanket surrounding the Earth is less. Therefore, the atmospheric pressure is less. At 5,000 feet elevation, the atmospheric pressure is only 12.2 pounds per square inch.

Absolute Pressure

Absolute pressure, normally expressed in terms of pounds per square inch absolute (psia) is defined as the pressure existing above a perfect vacuum. Therefore, in the air around us, absolute pressure and atmospheric pressure are the same.

Gauge Pressure

A pressure gauge is calibrated to read 0 pounds per square inch (psi) when it is not connected to a pressure-producing source. Therefore, the absolute pressure of a closed system will always be gauge pressure plus atmospheric pressure.

Pressures below 0 pounds per square inch gauge (psig) are actually negative readings on the gauge and are referred to as inches of vacuum. A refrigeration compound gauge (see Chapter 4) is calibrated in the

equivalent of inches of mercury for negative readings. Since 14.7 psi is
equivalent to 29.92 inches of mercury, 1 psi is approximately equal to
2 inches of mercury on the gauge dial.

It is important to remember that gauge pressures are only relative
to absolute pressure. Table 1.1 shows relationships existing at various
elevations, assuming that standard atmospheric conditions prevail.

The absolute pressure in inches of F mercury indicates the inches of
mercury vacuum that a perfect vacuum pump would be able to reach
(see Chapter 4). Therefore, at 5,000 feet elevation under standard atmospheric conditions, a perfect vacuum would be 24.89 inches of mercury.
as compared to 29.92 inches of mercury at sea level.

At very low pressures, it is necessary to use a smaller unit of
measurement, since even inches of mercury are too large for accurate
reading. The micron (μ), a metric unit of length, is used for this purpose;
and when we speak of microns in evacuation (see Chapter 4), we are
referring to absolute pressure in units of microns of mercury.

A micron is equal to 1/1000 of a millimeter, and there are 25.4
millimeters per inch. One micron, therefore, equals 1/25,400 inch.
Evacuation to 500 microns would be evacuating to an absolute pressure
of approximately .02 inch of mercury, or at standard conditions, the
equivalent of a vacuum reading of 29.90 inches of mercury.

TABLE 1.1   Pressure Relationships at Varying Altitudes

Altitude	Psig	Psia	Pressure in Inches Hg	Boiling Point of Water
0 ft.	0	14.7	29.92	212° F.
1000 ft.	0	14.2	28.85	210° F.
2000 ft.	0	13.7	27.82	208° F.
3000 ft.	0	13.2	26.81	206° F.
4000 ft.	0	12.7	25.84	205° F.
5000 ft.	0	12.2	24.89	203° F.

Courtesy Copeland Corporation

Pressure/Temperature Relationships: Liquids

The temperature at which a liquid boils is dependent on the pressure
being exerted on it. The *vapor pressure* of the liquid, which is the pressure being exerted by the tiny molecules seeking to escape the liquid
and become vapor, increases with an increase in temperature until—at
the point where the vapor pressure equals the external pressure—boiling
occurs.

Water at sea level boils at $212°$F but, at 5,000 feet elevation, it
boils at $203°$F, due to the decreased atmospheric pressure. If some

means, a compressor for example, were used to vary the pressure on the surface of the water in a closed container, the boiling point could be changed at will. At 100 psig, the boiling point is 327.8°F; and at 1 psig, the boiling point is 102°F.

Since all liquids react in the same fashion, although at different temperatures and pressures, pressure provides a means of regulating a refrigerating temperature. If a cooling coil is part of a closed system isolated from the atmosphere, and a pressure can be maintained in the coil equivalent to the saturation temperature (boiling point) of the liquid at the cooling temperature desired, then the liquid will boil at that temperature as long as it is absorbing heat. Refrigeration has been accomplished.

Pressure/Temperature Relationships: Gases

One of the basic fundamentals of thermodynamics is called the "perfect gas law." This describes the relationship of the three basic factors controlling the behavior of gas: pressure, volume, and temperature. For all practical purposes, air and highly superheated refrigerant gases may be considered perfect gases, and their behavior follows the following relation:

$$\frac{\text{Pressure}_1 \times \text{Volume}_1}{\text{Temperature}_1} = \frac{\text{Pressure}_2 \times \text{Volume}_2}{\text{Temperature}_2}$$

Although the perfect gas relationship is not exact, it provides a basis for approximating the effect on a gas of a change in one of the three factors. In this relation, both pressure and temperature must be expressed in absolute values, pressure in psia, and temperature in degrees Rankine or degrees Fahrenheit above absolute zero (°F plus 460°). Although not used in practical refrigeration work, the perfect gas relation is valuable for scientific calculations and is helpful in understanding the performance of a gas.

One of the problems of refrigeration is disposing of the heat that has been absorbed during the cooling process. A practical solution is achieved by raising the pressure of the gas so that the saturation or condensing temperature will be sufficiently above the temperature of the available cooling medium (air or water) to ensure efficient heat transfer. When the low-pressure gas with its low saturation temperature is drawn into the cylinder of a compressor, the volume of the gas is reduced by the stroke of the compressor piston, and the vapor is discharged as a high-pressure gas, readily condensed because of its high saturation temperature.

Specific Volume

The *specific volume* of a substance is defined as the number of cubic feet occupied by 1 pound of it. In the case of liquids and gases, it varies with the temperature and the pressure to which the fluid is subjected. Following the perfect gas law, the volume of a gas varies with both temperature and pressure. The volume of a liquid varies with temperature; but within the limits of practical refrigeration practice, it may be regarded as incompressible.

Density

The *density* of a substance is defined as weight per unit volume, and in the United States is normally expressed in pounds per cubic foot. Since by definition density is directly related to specific volume, the density of a gas may vary greatly with changes in pressure and temperature, although it still remains a gas that is invisible to the naked eye. Water vapor or steam at 50 psia pressure and 281°F temperature is over three times as heavy as steam at 14.7 psia pressure and 212°F.

1.5 PRESSURE AND FLUID HEAD

It is frequently necessary to know the pressure created by a column of liquid, or possibly the pressure required to force a column of refrigerant to flow a given vertical distance upwards.

Densities are usually available in terms of pounds per cubic foot, and it is convenient to visualize pressure in terms of a cube of liquid 1 foot high, 1 foot wide, and 1 foot deep. Since the base of this cube is 144 square inches, the average pressure in pounds per square inch is the weight of the liquid per cubic foot divided by 144. For example, since water weighs approximately 62.4 pounds per cubic foot, the pressure exerted by 1 foot of water is 62.4 divided by 144, which equals .433 pounds per square inch. Ten feet of water would exert a pressure of 10 times .433, which equals 4.33 pounds per square inch. The same relation of height to pressure holds true, no matter what the area of vertical liquid column. The pressure exerted by other liquids can be calculated in exactly the same manner if the density is known.

Fluid head is a general term used to designate any kind of pressure exerted by a fluid which can be expressed in terms of the height of a column of the given fluid. Hence, a pressure of 1 psi may be expressed as being equivalent to a head of 2.31 feet of water (1 psi divided by .433 psi/ft of water). In air flow through ducts, very small pressures are encountered, and these are commonly expressed in inches of water. One inch of water equals .433 divided by 12, which equals 0.36 psi (see Table 1.2).

TABLE 1.2   Pressure Equivalents in Fluid Head

Pounds per Square Inch	Inches Mercury	Inches Water	Feet Water
.036	.07	1.0	.083
.433	.80	12	1.0
.491	1.0	13.6	1.13
1.0	2.03	27.7	2.31
14.7	29.92	408	34.0

Courtesy Copeland Corporation

Fluid Flow

In order for a fluid to flow from one point to another, there must be a difference in pressure between the two points to cause the flow. With no pressure difference, no flow will occur. Fluids may be either liquids or gases, and the flow of each is important in refrigeration.

Fluid flow through pipes or tubing is governed by the pressure exerted on the fluid, the effect of gravity due to the vertical rise or fall of the pipe, restrictions in the pipe resisting flow, and the resistance of the fluid itself to flow.

For example, as a faucet is opened, the flow increases even though the pressure in the water main is constant and the outlet of the faucet has no restriction. Obviously, the restriction of the valve is affecting the rate of flow. Water flows more freely than molasses due to a property of fluids called *viscosity*, which describes the fluid's resistance to flow. In oils, the viscosity can be affected by temperature; as the temperature decreases, the viscosity increases.

As fluid flows through tubing, the contact of the fluid and the walls of the tube create friction and, therefore, resistance to flow. Sharp bends in the tubing, valves and fittings, and other obstructions also create resistance to flow, so the basic design of the piping system will determine the pressure required to obtain a given flow rate.

In a closed system containing tubing through which a fluid is flowing, the pressure difference between two given points will be determined by the velocity, viscosity, and the density of fluid flowing. If the flow is increased, the pressure difference will increase, since more friction will be created by the increased velocity of the fluid. The pressure difference is called *pressure loss* or *pressure drop.*

Since control of evaporating and condensing pressures is critical in mechanical refrigeration work, pressure drop through connecting lines can greatly affect the performance of the system; large pressure drops must be avoided.

Effect of Fluid Flow on Heat Transfer

Heat transfer from a fluid through a tube wall or through metal fins is greatly affected by the action of the fluid in contact with the metal surface. As a general rule, the greater this velocity of flow and the more turbulent the flow, the greater will be the rate of heat transfer. Rapid boiling of an evaporating liquid will also increase the rate of heat transfer. Quiet liquid flow, on the other hand, tends to allow an insulating film to form on the metal surface, which resists heat flow and reduces the rate of heat transfer.

1.6 REFRIGERANTS

Large quantities of heat can be absorbed by a substance through an increase in sensible heat involving either a big temperature difference or a large weight. In a change of state involving latent heat, however, a fraction of the weight will absorb an equivalent amount of heat.

In mechanical refrigeration, a process is required that can transfer large quantities of heat economically and efficiently, which can be repeated continuously. The processes of evaporation and condensation of a liquid are, therefore, logical components of the refrigeration process.

Practically any liquid could be used for absorbing heat by evaporation. Water is ideal in many respects, but it boils at temperatures too high for ordinary cooling purposes and freezes at temperatures too high for low temperature conditions. A refrigerant must satisfy two main requirements:

1. It must readily absorb heat at the temperature required by the product load.
2. For economy and continuous cooling, the system must use the same refrigerant over and over again.

Types of Refrigerants

There are many different types of refrigerant available, several of which are in common use. There is no perfect refrigerant, and there are varying opinions as to which may be best for specific applications. In early refrigeration applications, ammonia, sulfur dioxide, methyl chloride, propane, and ethane were widely used. However, due to the fact that they are either toxic, dangerous, or have other undesirable characteristics, they have been largely replaced in most applications by compounds developed especially for refrigeration use. Specialized refrigerants are used for ultra-low temperature work, or for large centrifugal compressors; but for normal commercial refrigeration, air conditioning, and heat-pump applications, refrigerants R-12, R-22, and R-502 are now used almost exclusively. Although these were developed originally by Dupont as Freon refrigerants, the numerical designations are now standard with all manufacturers.

Refrigerant 12 (R-12)

Refrigerant 12 is widely used in household and commercial refrigeration and air conditioning. At temperatures below its boiling point, it is a clear, almost colorless liquid. It is almost odorless, is not toxic or irritating, and is suitable for high, medium, and low temperature applications.

Refrigerant 22 (R-22)

Refrigerant 22, in most physical characteristics, is similar to R-12. However, it has much higher saturation pressures than R-12 for equivalent temperatures, has a much larger latent heat of evaporation, and a lower specific volume. As a result, for a given volume of saturated refrigerant vapor, R-22 has a much greater refrigerating capacity. This allows the use of lower compressor displacement, sometimes resulting in smaller compressors, for performance comparable with R-12. Where size and economy are critical factors, such as in packaged air conditioners and heat pumps, R-22 is widely used.

Refrigerant Saturation Temperature

At normal room temperatures, R-22 can exist only as a gas unless under pressure, since its boiling point at atmospheric pressure is far below $0°F$. Therefore, refrigerants are always stored and transported in special pressure resistant drums. As long as both liquid and vapor are present in a closed system with no external pressure influence, the refrigerant will either evaporate or condense, depending on the outside temperature, until the saturation pressure and temperature corresponding to the outside temperature is reached and heat transfer can no longer take place. A decrease in outside temperature will allow heat to flow out of the refrigerant, cause condensation, and lower the pressure; an increase in outside temperature will cause heat to flow into the refrigerant, cause evaporation, and raise the pressure.

Refrigerant Evaporation

Presume the refrigerant is enclosed in a refrigeration system, with its temperature equalized with the outside temperature. Instead of changing with the outside temperature, the pressure in the refrigeration system is lowered. Since this lowers the saturation point, the temperature of the liquid refrigerant is now above its boiling point. It will immediately start boiling violently, absorbing heat in the process, and thus reducing the temperature of the remaining liquid and changing into gas as the change of state takes place.

Heat will now flow into the system from the outside due to the decreased temperature of the refrigerant. Boiling will continue until the outside temperature is reduced to the saturation temperature of the refrigerant, or until the pressure in the system again rises to the equivalent saturation pressure of the outside temperature. If a means is provided, such as a compressor, to remove the refrigerant vapor so the pressure does not increase, while at the same time liquid refrigerant is fed back into the system, continuous refrigeration will be taking place. This is basically the process taking place in a refrigeration-system evaporator.

Refrigerant Condensation

Again, presume the refrigerant is enclosed in a refrigeration system, with its temperature equalized with the outside temperature. If hot refrigerant vapor is pumped into the system, the pressure in the refrigeration system is increased, thus raising the saturation point.

As heat is transferred from the incoming hot vapor to the refrigerant liquid and the walls of the system, the temperature of the refrigerant vapor falls to its condensing temperature and condensation starts. Heat from the latent heat of condensation flows from the system to the outside until the pressure in the system is lowered to the equivalent of the saturation pressure at the outside temperature. If a means is provided, such as a compressor, to maintain a supply of hot, high-pressure refrigerant gas, while at the same time liquid refrigerant is drawn off, continuous condensation will take place. This is basically the process taking place in a refrigeration-system condenser.

Refrigerant/Oil Relationships

In compressors, oil and refrigerant mix continuously. Refrigeration oils are soluble in liquid refrigerant, and at normal room temperatures they will mix completely. The ability of a liquid refrigerant to mix with oil is termed *miscibility*, and the refrigerant is described as being miscible with oil.

Oil circulating in a refrigeration system may be exposed to both very high and very low temperatures. Because of the critical nature of lubrication under these conditions and the damage that can be done to the system by wax or other impurities in the oil, only highly refined oils specially prepared for refrigeration usage can be used.

In general, naphthenic oils are more soluble in refrigerants than paraffinic oils. Separation of the oil and refrigerant into separate layers can take place with both types of oil, although at somewhat lower temperatures with naphthenic oils. Separation does not necessarily affect the lubricating ability of the oil, but it may create problems in properly supplying oil to the working parts.

Since oil must pass through the compressor cylinders to provide lubrication, a small amount of oil is always circulating with the refrigerant. Oil and refrigerant gas do not mix readily, and the oil can be properly circulated through the system only if gas velocities are high enough to sweep the oil along. If velocities are not sufficiently high, oil will tend to lie on the bottom of the refrigerant tubing, decreasing heat transfer and possibly causing a shortage of oil in the compressor. As evaporating temperatures are lowered, this problem becomes more critical, since the viscosity of the oil increases with a decrease in temperature. For these reasons, proper design of piping is essential for satisfactory oil return.

One of the basic characteristics of a refrigerant and oil mixture in a sealed system is the fact that refrigerant is attracted by oil and will vaporize and migrate through the system to the compressor crankcase, even though no pressure difference exists to cause the movement. On reaching the crankcase, the refrigerant will condense into liquid, and this migration will continue until the oil is saturated with liquid refrigerant.

Excess refrigerant in the compressor crankcase can result in violent foaming and boiling action, driving all of the oil from the crankcase and causing lubrication problems. Therefore, provisions must be made to prevent the accumulation of excess liquid refrigerant in the compresssor. This will be discussed in detail in Chapter 3.

1.7 A TYPICAL REFRIGERATION CYCLE

Let's now look at a typical refrigeration cycle (shown in Figure 1.2). There are two pressures existing in a compression system, the evaporating or low pressure, and the condensing or high pressure. The refrigerant acts as a transportation medium to move heat from the evaporator to the condenser, where it is given off to the ambient air, or in a water-cooled system, to the cooling water. A change of state from liquid to

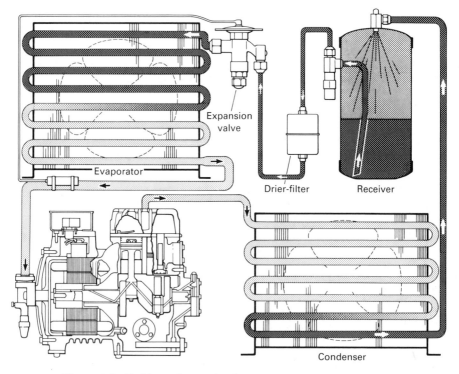

Figure 1.2   Refrigeration cycle (Courtesy Copeland Corporation).

vapor and back to liquid allows the refrigerant to absorb and discharge large quantities of heat efficiently.

The basic cycle operates as follows:

High-pressure liquid refrigerant is fed from the receiver through the liquid line, and through the filter-drier to the metering device separating the high-pressure side of the system from the low-pressure evaporator. Various types of control devices may be used, but for purposes of this illustration, only the thermostatic expansion valve will be considered.

The thermostatic expansion valve controls the feel of liquid refrigerant to the evaporator and, by means of an orifice, reduces the pressure of the refrigerant to the evaporating or low-side pressure. The reduction of pressure on the liquid refrigerant causes it to boil or vaporize until the refrigerant is at the saturation temperature corresponding to its pressure. As the low-temperature refrigerant passes through the evaporator coil, heat flows through the walls of the evaporator tubing to the refrigerant, causing the boiling action to continue until the refrigerant is completely vaporized.

The expansion valve regulates the flow through the evaporator as necessary to maintain a preset temperature difference or superheat between the evaporating refrigerant and the vapor leaving the evaporator. As the temperature of the gas leaving the evaporator varies, the expansion valve power element bulb senses its temperature and acts to modulate the feed through the expansion valve as required.

The refrigerant leaving the evaporator travels through the suction line to the compressor inlet. The compressor takes the low-pressure vapor and compresses it, increasing both the pressure and the temperature. The hot, high-pressure gas is forced out through the compressor discharge valve and into the condenser.

As the high-pressure gas passes through the condenser, it is cooled by some external means. In air-cooled systems, a fan and fin-type condenser surface is normally used. In water-cooled systems, a refrigerant-to-water heat exchanger is usually employed. As the temperature of the refrigerant vapor reaches the saturation temperature corresponding to the high pressure in the condenser, the vapor condenses into a liquid and flows back to the receiver to repeat the cycle. This refrigeration process will remain continuous as long as the compressor operates.

1.8 A TYPICAL HEAT PUMP CYCLE

Let's now look at a typical heat pump cycle, which utilizes the same principles discussed with regard to refrigeration; i.e., heat is continually moved from a region of low temperature to a region of high temperature due to the use of a refrigerant.

The simplest way to explain a heat pump cycle is to trace the flow of refrigerant as it travels through the heat pump system. However, since a heat pump is capable of both heating and cooling, which is accomplished through a reversal of the flow of refrigerant within a closed system, the heating and cooling cycles will be discussed individually. Also, although the heat pump, like refrigeration equipment, has two coils, the coils reverse their actions (evaporation and condensation) depending on cycle. For this reason, it would be confusing to label them as evaporator coil and condenser coil and then reverse these labels when a shift in cycle occurs. In a heat pump, then, the coils are normally referred to by their place within the system. The outdoor coil is in the outdoor air stream, while the indoor coil is in the indoor air stream. Also, for purposes of this discussion, the heating and cooling cycles of an air-source heat pump will be described.

Let's first trace the flow of refrigerant through an air-source heat pump during the cooling cycle, when the system is removing heat from the conditioned space and rejecting it outdoors. Starting at the compressor discharge, the hot refrigerant gas is discharged to the reversing valve, which directs it to the outdoor coil (Figure 1.3). The outdoor

COOLING CYCLE

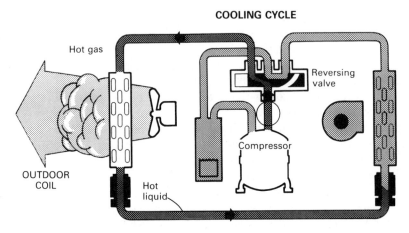

Figure 1.3 Hot refrigerant gas is directed to the outdoor coil by the reversing valve (Courtesy Carrier Corporation).

coil acts as a condenser. Outdoor air circulated over the coil removes heat from the refrigerant gas and causes it to condense to a hot liquid.

The liquid refrigerant then passes through a metering device located near the indoor coil (Figure 1.4) where the pressure is reduced, causing some of the liquid to flash off and cool the remaining liquid to a lower temperature. This cool liquid passes through the indoor coil, which acts as an evaporator. The liquid boils and absorbs heat from the indoor air passing over the coil, causing the refrigerant to evaporate into a cool vapor.

COOLING CYCLE

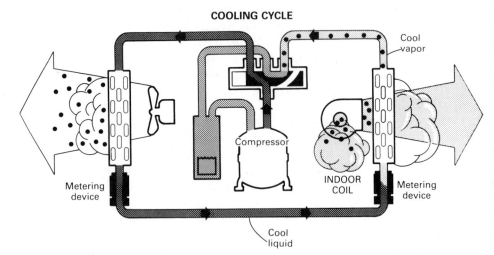

Figure 1.4 Cool liquid refrigerant is directed to the indoor coil (Courtesy Carrier Corporation).

The cool vapor passes through the reversing valve, which directs it to the accumulator, where any liquid refrigerant will be trapped, and then back to the compressor, where the gas is compressed to a high-pressure, high temperature gas, and the cycle is repeated (see Figure 1.5).

During the heating cycle when heat is removed from the outside air and rejected indoors, the hot-gas discharge of the compressor is directed by the reversing valve to the indoor coil (Figure 1.6). The indoor coil now becomes a condenser. The indoor air passes over the coil and picks up heat from the warm refrigerant gas passing through the coil. This causes the refrigerant gas to condense to a warm liquid.

COOLING CYCLE

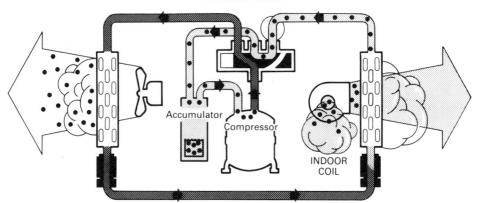

Figure 1.5 Cool vapor is directed back to the compressor, and the cycle is repeated (Courtesy Carrier Corporation).

HEATING CYCLE

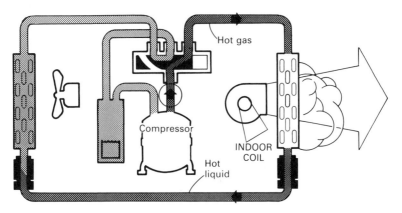

Figure 1.6 Hot refrigerant gas is directed to the indoor coil by the reversing valve (Courtesy Carrier Corporation).

HEATING CYCLE

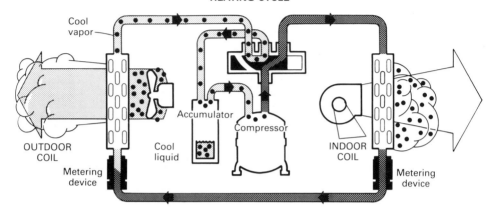

Figure 1.7 Warm liquid refrigerant is directed to the outdoor coil (Courtesy Carrier Corporation).

The liquid refrigerant flows to the outdoor coil where it passes through a metering device and flashes to a low-pressure, low temperature mixture of liquid and vapor in the outdoor coil (Figure 1.7). The outdoor coil acts as an evaporator. Heat from the outdoor air is absorbed into the refrigerant as it boils into a cool vapor. The refrigerant gas goes through the reversing valve, which directs it to the accumulator and back to the compressor, where it is compressed and the cycle is repeated.

From this discussion, it can be seen that heat is in fact moved by the heat pump. The total amount of heat rejected by the coil (which is acting as a condenser) is equal to the total heat absorbed into the system plus the heat of compression.

This chapter has presented the principles that are used to create a system in which both heating and cooling can take place by controlling the flow of a refrigerant. Later chapters will describe the different types of heat pumps, their internal components and circuits, and the installation, troubleshooting, and repair of heat pumps.

Two

Introduction
to Heat Pumps

In this chapter, we will learn about the history of heat pumps: how and why they came to be designed, their efficiency as compared to conventional heating systems, and the different types of heat pumps in use today.

2.1 A BRIEF HISTORY

It is a common misconception that the heat pump is a relatively new development. The basic principle of heat pump operation was first proposed by Nicholas Carnot in 1824 and then furthered by Lord Kelvin in 1852. Lord Kelvin proposed a system capable of both heating and cooling, in which a compressor would be used as a means of providing indirect heat, instead of equipment that burned a combustible fuel directly. Several early model heat pumps were constructed (utilizing water as the heat source), but proved unworkable with the components available at the time. The reversing action needed to provide heating or cooling on demand was carried out manually, and required the services of experienced mechanics to open or close the dozen or so valves that allowed for the shift from heating to cooling and then back again.

In the late 1940s, some progress was made in equipment development. Rather than utilizing water as the heat source, most manufacturers put their energies into air-source heat pumps, because many designers were under the mistaken impression that ground water resources were limited. One air-source heat pump manufactured during this period reversed air flow over the coils rather than reversing the flow of the refrigerant. In other words, when called upon for cooling, the indoor air was directed over the indoor coil; when called upon for heating, the indoor air was directed over the outdoor coil. The equipment, however, was quite large and extremely undependable.

At the same time, air conditioning equipment was gaining in popularity and efficiency; by the time World War II was over, there were so many air conditioners being installed that power companies were experiencing higher power loads in summer than in winter. In an effort to equalize power loads, power companies began an active promotional campaign in support of heat pumps, in spite of the fact that the equipment was still not perfected.

The '50s heralded further advances in heat pump technology, but since fossil fuels were so inexpensive and readily available, the heat pump remained a relatively obscure concept. Although heat pumps were commercially manufactured in 1952, the systems were still quite inefficient and unreliable.

The Arab oil embargo of the early 1970s finally brought home the message to the American people that alternative energy sources warranted serious consideration. As fuel costs continued to rise, many persons were forced to consider other means of heating their homes, and manufacturers began investing large amounts of money into the research and development of the heat pump. This research and development, which took place over a decade, has resulted in the manufacture of heat pump equipment that is considered to be reliable and efficient, and more importantly, costs considerably less to operate than conventional heating systems. Although research continues today in an effort to further improve efficiency, the general consensus is that the heat pumps being manufactured for today's market are a significant achievement. The equipment is cost effective, reliable, and a viable alternative to conventional heating and cooling methods.

2.2 THE EFFICIENCY OF A HEAT PUMP

Most homes today utilize natural gas, fuel oil, or electricity to heat their homes, and electricity to provide hot water. Until the early 1970s, these heating methods, although inefficient, were considered extremely cost effective. Prior to the Arab oil embargo, fuel oil was about 35 cents a gallon and electricity was about 2 cents a kilowatt hour. Today, the same gallon of fuel oil is costing upwards of a dollar, and the same kilowatt hour of electricity is costing upwards of 6 cents.

At the same time, many homeowners who purchased homes 20 or 30 years ago, when fuel costs were low, are faced with the prospect of extensive repairs to an existing heating system that is near the end of its useful life. Those persons considering the building or purchase of a new home are also faced with the cost of installing a heating and cooling system, and must take into consideration not only the installation costs but the operating costs of such a system over its useful life.

The heat pump is a viable alternative in a new or existing home. It has been proven to be more cost effective in the majority of installations, even when installation costs are considered. A heat pump can be installed in conjunction with an existing heating system (with some modifications to the system) to take over the majority of the work involved in heating the home. The existing furnace will become a "back-up" system, which is only used when temperatures fall below the point where the operation of the heat pump is efficient, or when the heat pump breaks down and needs servicing. The amount of time that an existing system is utilized will depend upon ambient temperatures in the area of the country where the system is installed and the type of heat pump used, but costs will still be much lower than if a conventional system were used alone.

It is difficult to give actual figures for measuring a heat pump's efficiency as compared to conventional heating, because fuel oil, propane, and electric costs vary throughout the country, as do ambient temperatures, installation costs, and other variables that are applied in such measurements. Generally, when the installation of a heat pump is under consideration, the installer will perform a series of calculations that will compare the home's existing system with that of a heat pump system. These calculations will take into consideration the cost of installation, energy costs, temperature ranges, and a number of other variables. The homeowner is then able to see whether a heat pump is cost effective. The heat pump is usually significantly more cost effective than any other type of heating system.

It should be apparent, from the information presented in Chapter 1, that the heat pump moves heat. It does not produce heat directly. For illustration purposes, let's compare the efficiency of a heat pump with electric heat. This can be measured using one of two rating methods.

One method is the energy efficiency ratio (EER) that is used to rate any air conditioning system. Another method that has been used for many years when discussing the efficiency of a heat pump is the coefficient of performance (COP). No matter which method is used, a knowledge of how they are determined is important.

The energy efficiency ratio is nothing more than how many BTUs that are produced for every watt of power used. It is determined by dividing the total capacity (in BTU/hr) of the system by the total electric power consumed in watts per hour. For example, if a 3-ton heat pump operating on a 40°F outdoor temperature has a heating capacity of 39,000 BTU/hr and is consuming 4,380 watts per hour, the EER would be 8.9 BTU/watt (39,000 divided by 4,380 = 8.9).

To determine the COP of a heat pump, this formula can be used:

$$COP = \frac{BTUs\ out}{BTUs\ we\ pay\ for}$$

or

$$\frac{BTU/hr\ capacity}{Unit\ wattage \times 3.413\ BTU/watt}$$

If we use the same 3-ton unit as in our last example, the COP would be:

$$\frac{39,000\ BTU/hr}{4,380\ watt/hr \times 3.413\ BTU/watt} = 2.6$$

The COP indicates how efficient the unit is when compared with electric resistance heat. When electric heat is used, it generates 3,413 BTUs of heat for each watt. Therefore, the COP of straight electric heat is one (1.0) and can never be any higher. When comparing the COP of the heat pump with that of an electric heater, we find that the heat pump can deliver more heat per watt of power used. The heat pump is, therefore, more efficient. Under the operating conditions in our example, the heat pump will deliver 2.6 times as much heat as an electric resistance heater using the same wattage.

A fair method of comparing the heat pump to other heating systems is to determine the annual COP, which is the average performance of the heat pump throughout the entire heating season. This annual COP rating is called the seasonal performance factor (SPF).

2.3 TYPES OF HEAT PUMPS

Heat pumps are classified by the heat source used during the heating cycle. The two major classifications are air-source heat pumps and water-source heat pumps. Within these classifications, there are further classifications as to the type of packaging and the medium to which the heat is transferred. The basic operating principles, however, remain the same regardless of the heat source, although some components will vary. Each type will be discussed individually.

Air-source Heat Pumps

The air-source heat pump, as its name implies, utilizes the outside air as a heat source. Because the outside air is free and readily available, the air-source heat pump is by far the more popular of the two types of heat pumps, although there are inherent disadvantages associated with its dependence on the outside air.

Since the air-source heat pump must have access to the outside air, all or a portion of the heat pump system must be located outdoors. This can be accomplished in two ways. The components and circuitry can be housed in a single unit that is installed outdoors, or the components and circuitry can be housed in two units, one installed outdoors and one installed indoors. A heat pump that is contained in a single unit is called a *packaged heat pump*, while a heat pump that is contained in two units is called a *split-system heat pump* (Figure 2.1). Both packaged and split systems can be either horizontal or vertical as well. If a heat pump is installed in conjunction with an existing heating system, whereby the heat pump becomes the primary source of heat and the heating system becomes the secondary source of heat, the heat pump is called an *add-on split system*. An add-on system always requires two units, and the only difference between a split system and an add-on system is that the indoor unit becomes an integral part of the existing heating system in the latter case.

Figure 2.1   The Concept 1 split-system heat pump (Courtesy Magic Chef Air Conditioning).

An air-source heat pump can extract heat from the outside air when temperatures are quite low because heat is always present to some degree. Even at $0°F$, the outside air contains 89% of the heat available at $100°F$. In fact, there is some heat available in the air down to $460°$ below $0°F$. As long as the refrigerant in the heat pump's outdoor coil is at a lower temperature than the outside air, heat will move into the refrigerant. However, even though the heat pump is able to extract heat from the outside air at extremely low temperatures, it must work harder to do so; this is the major disadvantage of an air source heat pump. The lower the temperature, the harder the heat pump must work, and the less heat it is able to extract from the outside air.

When the outside temperature reaches the point where the heating capacity of the heat pump equals the heating requirements of the home (called the *balance point*), it becomes more efficient to utilize a supplementary form of heat (electric resistance units installed within the heat pump or a conventional heating system acting as a back-up system). The amount of time that a supplemental form of heat is required will depend on ambient temperatures in the area where the heat pump is installed. However, almost all of the air-source heat pumps installed today are installed with supplemental heat in some form or another.

As discussed in Chapter 1, there are two major cycles of operation in a heat pump: heating and cooling. An air-source heat pump, however, requires the components and circuitry necessary for a third cycle, the defrost cycle.

It is quite possible for the outdoor coil temperature to be below freezing when a heat pump is in the heating cycle. Under these conditions, any moisture removed from the air passing over the outdoor coil will freeze on the surface of the cold coil. Eventually, the frost on the coil can build enough to reduce the amount of air passing over the coil and the coil will lose its efficiency. Frost will also insulate the finned surface and cut down on heat transfer. This decreases coil efficiency even further. When the coil efficiency is sufficiently impaired to appreciably affect the system capacity, the frost must be removed. This is done by shifting the reversing valve into the cooling cycle, which directs the hot discharge gas to the outdoor coil to melt the frost.

Because the only time that frost can build up on the outdoor coil is during winter, and because frost removal requires a shift from heating to cooling in order to direct the hot refrigerant to the outdoor coil, the second disadvantage of the air-source heat pump should now be apparent. Even if supplementary heat were not required due to cold ambient air temperatures, it is required when frost builds up on the outdoor coil during the winter. Supplemental heat is a necessity in an air-source heat pump, as is the additional cost of providing defrost circuitry and components.

Water-source Heat Pumps

The water-source heat pump utilizes water as the heat source. This can be accomplished in a number of ways, all of which involve delivering a supply of water to the outdoor coil of the heat pump. Unlike the air-source heat pump, no portion of a water-source heat pump need be located outdoors; most water-source heat pumps (with the exception of add-on systems) are packaged systems. Also, because the outdoor coil is not exposed to the outside air, there is no danger of frost build up, and thus, no need for defrost circuitry and components.

The water-source heat pump has been the focus of considerable controversy, and it is only in recent years that much of this controversy has been dispelled. As mentioned in Chapter 1, early heat pumps utilized a water source in some form, but because many persons were under the mistaken impression that ground water supplies were limited, manufacturers turned their energies toward mass production of air-source heat pumps instead. However, because of the inherent disadvantages and additional costs associated with utilizing a heat source whose temperature is constantly changing, some manufacturers began intensive programs to study the feasibility of a heat pump that utilized water as the heat source. These studies have shown conclusively that the water-source heat pump is a viable alternative to the air-source heat pump if adequate water can be supplied. In fact, the water-source heat pump has been proven to be considerably more efficient than its air-source counterpart.

Water-source heat pumps are more efficient than air-source heat pumps because they are extracting heat from water that is maintained at a near constant temperature year-round. Underground water is kept at a near-constant temperature, because the sun's heat is continually being absorbed into the ground through the earth's surface. It is estimated that at depths of 10 feet or more, soil temperatures throughout the United States are maintained at a fairly constant $42°$ to $45°F$, which results in a near-constant temperature underground water supply. Because the water-source heat pump is supplied with a near-constant temperature water supply, the system has a high coefficient of performance (COP). The water-source heat pump does not have to work harder in the winter to obtain heat, as does the air-source heat pump; because the water temperature remains nearly the same, no supplemental form of heat is necessary. Supplemental heat, however, is often provided with a water-source heat pump as an emergency back-up system in the event of system failure.

There are two major types of water-source heat pumps in use today. In each system, the water is supplied to the heat pump in a different manner. These are the *ground-water heat pump* and the *closed-*

loop heat pump. The ground-water heat pump utilizes a water-well system, while the closed-loop heat pump utilizes a closed loop piping that is buried underground and then filled with water (Figure 2.2). Either method of supplying water to the heat pump will, of course, create additional installation costs; this is one reason why water-source heat pump installations are far fewer than air-source heat pumps. However, once a water-source heat pump has been installed, it will cost considerably less to operate than an air-source heat pump. It is estimated that payback of installation costs can occur in 3 to 4 years. Another reason why installations of air-source heat pumps far exceed water-source heat pumps is that space limitations, particularly in urban areas, limits the use of the latter system. This will be discussed in detail later in this chapter.

Although the manner in which water is supplied in each type of water-source heat pump is unique, the basic operating principles already discussed with regard to air-source heat pumps remain the same. In the winter (Figure 2.3), the water is delivered to the heat pump's outdoor coil (more often referred to as a *heat exchanger* or *water coil*), where heat is absorbed by the high-temperature, high-pressure refrigerant and then transferred to the space to be heated. In the summer (Figure 2.4), the principle is reversed, and heat from inside of the home is absorbed by the low-temperature, low-pressure refrigerant in the indoor heat

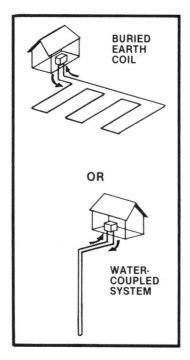

BURIED
EARTH
COIL

OR

WATER-
COUPLED
SYSTEM

Figure 2.2   Earth-coupled or Water-coupled systems (Courtesy National Well Water Association).

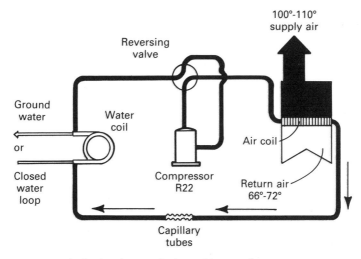

In the heating mode, hot refrigerant flows
through the air coil supplying warm air to the
conditioned space.

Figure 2.3   The heating cycle of a water-source heat pump (Courtesy
National Well Water Association).

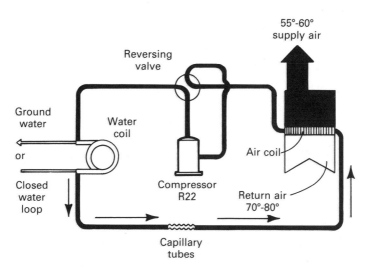

In the cooling mode, cold refrigerant flows
through the air coil supplying cool air to the
conditioned space.

Figure 2.4   The cooling cycle of a water-source heat pump (Courtesy
National Well Water Association).

exchanger, causing its temperature to increase. The refrigerant then travels to the water coil, where the heat is transferred to the water. The same cycles of evaporation and condensation take place, with the only difference being the heat source and the manner in which the heat, or thermal energy, is retrieved.

The major physical difference between air- and water-source heat pumps is the outdoor coil or heat exchanger. As shown in Figure 2.5, a tube-in-tube type of construction is used whereby the water and refrigerant travel in opposite directions through the coil. The water either absorbs heat from the refrigerant or delivers heat to the refrigerant, depending on the cycle. Most other components in water-source heat pumps, regardless of type, are very similar and exhibit the same basic characteristics.

Ground-water Heat Pumps

A ground-water heat pump absorbs heat from ground water (in the form of a supply well) during the cooling cycle, and discharges heat to the ground water (in the form of a discharge well) during the heating cycle (Figure 2.6). An existing well may be used, but it must exhibit the proper characteristics (adequate supply, good quality, and so on), and be able to meet both the needs of the heat pump and the home as well. Also, even if an existing well is utilized, a satisfactory means of delivering the discharge water back to the ground must be available,

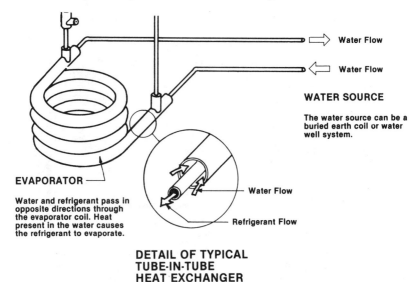

Figure 2.5   Water-coil (Courtesy National Well Water Association).

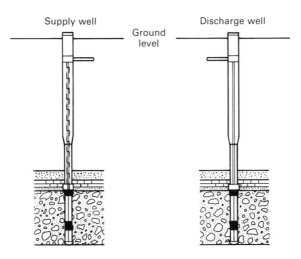

Figure 2.6 Supply and discharge wells (Courtesy Command-Aire Corporation; Waco, Texas).

usually in the form of a discharge well. More often than not, a new supply and discharge well are necessary for proper operation of a ground water heat pump.

The National Well Water Association defines ground water in this way:

> Ground water is the water which fills pores and cracks in underground rocks and soil. Like petroleum, uranium or cobalt, ground water has economic value. But unlike other economic resources and minerals, ground water is part of the endless hydrologic cycle. Over 93% of the water supply in the United States lies underground. Yet, most of our current water supply comes from surface sources such as lakes, rivers and streams.
>
> Ground water is replenished by nature depending on the local climate and geology and is variable in both amount and quality. When rain falls, the plants and soils take up water. Some of the excess water runs off to streams, and some percolates down into the pores and cracks of the subsurface rocks. Ground water does not flow in veins, domes or underground rivers or lakes. A water well that extends into the saturated zone will fill with water to the level of the water table—the top of the zone in which all of the openings of the rocks are filled with water (see Figure 2.7).[1]

In 1976, the National Well Water Association undertook an extensive study to determine if ground water resources were limited, as many within the industry thought. They concluded that over 85% of the con-

---

[1] *Understanding Heat Pumps, Ground Water and Wells* (National Well Water Association, 1983), p. 2.

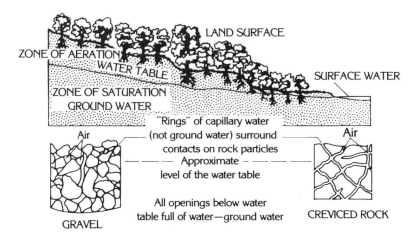

Figure 2.7 How water occurs in rocks (Courtesy National Well Water Association).

tinental United States had sufficient shallow ground water resources to use the ground-water heat pump for residential heating and cooling:

> While ground water receives a continuous supply of geothermal energy from the earth's interior, it is also a natural solar storage facility. Heat energy from the sun is absorbed by surface water which moves through the earth's crust to the water table. This energy is stored in shallow ground water *aquifers*, bodies of rocks that contain usable water supplies.[2]

The temperature of available ground water is maintained by a combination of the sun's energy, geothermal energy from the earth's interior, and location. The temperature can range from approximately $45°F$ in the north to approximately $80°F$ in the extreme south. Studies have indicated that a ground-water heat pump is able to operate efficiently with ground water temperatures as low as $39°F$. Obviously, the higher the temperature of the ground water, the more efficient the operation of the heat pump.

A ground-water heat pump does not consume water; rather, the same amount of water that travels to the heat pump travels back out. For this reason, a satisfactory means of discharging the water is required. From a water conservation standpoint, the preferred method of discharge is to return the water to the aquifer from which it was received, usually in the form of a return well. The return well is drilled to empty into the same aquifer as the supply well, with the return well being drilled far enough from the supply well to prevent an overlapping ther-

[2]*Understanding Heat Pumps*, p. 4.

mal effect between the two wells. The temperature of the water traveling out of the heat pump is higher than the water traveling into the heat pump during the cooling cycle, because heat is transferred from the home to the water. Generally, the supply and return wells are spaced at least 100 feet apart, although this may vary depending on the heat pump, the area of the country in which the system is installed, and the characteristics of the aquifer itself. In cases where the discharge water is returned to an aquifer other than the supply aquifer, thermal interference is negligible or nonexistent. However, supply and return aquifers are normally tested to ensure that they are chemically compatible, so that mixing of two water types does not occur and result in plugging the aquifers.

The heat pump installer will require the services of a certified well driller to ensure that the proper quality, quantity, flow, and discharge of water is provided. The flow requirement will depend on the particular heat pump being installed. As an example, a 50,000 BTU heat pump can require from 5 to 15 gallons per minute (gpm), or 7,200 to 21,600 gallons per day. In some areas of the country, there are stringent regulations that must be adhered to regarding the quantity of ground water that may be used, and it will normally be the responsibility of a certified well driller to assure that the system conforms with any regulations that may apply.

A ground-water heat pump will also require the installation of a well pump to deliver water to the system. The well driller will select the proper pump, considering many factors, including the demand of the system in gallons per minute and the depth of the well. He will also normally install all outdoor piping from the well to the heat pump.

Generally, the piping is sized for the required water flow and the lowest friction loss to maximize efficiency. Supply lines from the well to the home are normally buried below the frost line to prevent freezing, although they are sometimes insulated as insurance against heat loss and decreased efficiency. This will depend on the area of the country in which the system is installed. Return lines are not normally insulated, although they are also buried below the frost line. Supply and return lines are installed far enough away from each other to prevent any heat exchange from taking place.

Ground-water heat pumps are considerably more efficient than air-source heat pumps, although they pose higher installation costs. The efficiency of a heat pump, as you know, is measured by the number of units of heat-energy output in BTUs obtained for each unit of heat-energy input in kilowatts (the coefficient of performance or COP).

The COP is a measure of heat pump efficiency; the higher the COP, the more efficient the heat pump. The COP of a heat pump is more than 1 because the heat energy exceeds the input electrical energy. Conventional heating systems generally have lower accepted COPs than ground-water heat pump systems. Some average seasonal COPs of those systems under ideal circumstances are listed below:

Electrical resistance furnace   1.0

Natural gas furnace   0.80

Coal furnace   0.60

Fuel oil furnace   0.70

In other words, a conventional natural gas furnace system produces 70 BTUs of heat for every 100 BTUs of energy it consumes, or has an efficiency of about 70% if properly maintained.

Heat pumps, on the other hand, normally have COPs greater than 1. An air-source heat pump typically has a COP of 1.5 to 2.0. The fact that heat pumps typically have COPs greater than 1 seems to contradict the laws of nature, since more energy comes out of the heat pump than is put in. Actually, a heat pump merely uses electrical energy to transfer free thermal energy from the outside air or from a water source.

Ground water has been called nature's most efficient energy storage system. In addition to the advantage of high specific heat, ground water's temperature varies little with the seasonal weather changes. It is warmer than the air in winter and cooler than the air in summer. Therefore, the heat pump output is also relatively constant year-round. It is for this same reason that the performance of the air-source heat pump varies more than that of a ground-water system, because the temperature of the air source is dependent upon outside weather conditions. When air-source heat pumps operate in extreme temperatures, they consume large quantities of electricity in order to operate effectively and often must rely upon a back-up heating method during the winter (such as electrical resistance strip heating) to provide a comfortable home temperature. Ground-water heat pumps under operating conditions are about 25%–75% more efficient than air-source heat pumps operating under optimal conditions.

Here's an example of the efficiency of the ground-water heat pump. An average house of 2,000 square feet of living area requires about 360,000 BTUs per day for heating. It requires about 28.8 kilowatt hours (kwh) per day to heat the home to $68°$ F using a ground-water heat pump with $45°$ F ground water and an outside air temperature of $20°$ F. The usual energy requirement for an electrical resistance heating system under similar conditions would be 108 kwh per day. For every kilowatt hour of energy put into the system, you would be getting 3.6 times the BTUs per hour out. Thus, the ground-water heat pump's COP is 3.6.

In the cooling mode, ground-water heat pumps are often rated on the basis of an energy efficiency ratio (EER). EER represents output divided by power input; a higher number indicates better efficiency. EER ratings for air-source heat pumps fall between 6.8 and 9.0. Ground-water heat pumps can reach EER ratings of 13 or more.

Table 2.1 compares the COPs of the various systems and shows the rela-

TABLE 2.1   Comparison of the COPs Various Heating Systems

System	COP	Energy Input	Energy Output
Electrical resistance	1.0	100	100
Fuel oil	.70	100	70
Propane	.75	100	75
Natural gas	.80	100	80

*Computer simulations conducted for the U.S. Department of Energy study, "Ground Water Heat Pumps: An Examination of Hydrogeologic, Environmental, Legal and Economic Factors Affecting Their Use," determined that the ground water heat pump uses 20 to 60 percent less energy for heating than the air-source heat pump.

(Courtesy National Well Water Association)

tionship between the energy input (what you buy) and the energy output of the system. The energy output of the system is the energy input multiplied by the COP. Note that the average COP of a ground-water heat pump is significantly higher than the air-source heat pump.

The higher the COP, the lower the annual operating cost of the system will be. In comparison, 100 BTUs of energy will produce 320 BTUs of heat from a ground-water heat pump, but only 75 BTUs from a natural gas furnace. This reduction in annual operating costs offsets the higher initial costs of a ground-water system.

A computer simulation by the National Well Water Association of currently available ground-water heat pumps located in nine cities through the United States predicts annual COPs ranging from 2.2 in Concord, New Hampshire (where 94% of the energy load was for heating), to 2.9 in Birmingham, Alabama (where approximately 50% was for heating and 50% for cooling). In Concord, only 10% of the total energy consumed by the ground-water heat pump system was used for supplemental electric strip heat, whereas an air-source heat pump required 40% of its energy consumption for the same purpose. By simulating a heat-only ground-water heat pump and using direct heat-exchange cooling (with ground water), energy requirements were reduced 30% compared to the reversible-cycle ground-water heat pump. This is a result of sizing to full heating design load rather than cooling load. With this strategy, the use of supplemental electric strip heat can be substantially reduced or eliminated. However, some suggest that the use of supplemental electric strip heat, which will probably only be needed for a few days a year, should be less expensive than greatly increasing the size of the heat pump system.[3]

[3]*Understanding Heat Pumps*, pp. 14–15.

As mentioned earlier, many homeowners shy away from ground-water heat pumps due to higher initial costs. However, even though the initial cost of a ground-water heat pump is higher than that of conventional systems, it exhibits lower operating costs, decreased maintenance, and the ability to provide both heating and cooling. Of course, costs will vary somewhat depending upon the area of the country where the system is installed, the condition of the home, the temperature of the ground-water supply, the type of equipment, the water discharge method selected, and other variables such as interest rates and tax incentives:

> Based on U.S. Department of Energy projections of energy costs, a gas heating/cooling system is the most economically attractive of current system choices in most parts of the United States. However, the ground-water heat pump, with no well costs included, has an economic advantage over all other systems evaluated (including air-source heat pumps, electrical resistance heating/electrical cooling systems and oil heating/electrical cooling systems) in eight of the nine test cities of the study, with Houston, Texas being the exception.
>
> With the cost of a return well included, payback of incremental first costs for installation of a ground-water heat pump system is usually achieved within an 8-year life cycle period. The shortest payback periods are indicated for northern climate installation. Using the air-source heat pump or the electric heating/electric cooling system as the alternate choice, payback periods range from 2 to 10 years. Using the oil heating/electric cooling system as the alternate, payback periods range from 1 to more than 10 years.[4]

The total cost for the installation of a residential ground-water heat pump system, including well drilling, can range from approximately $5,600 to $8,400, which is considerably more than an air-source system, which usually costs approximately $3,500. A conventional heating system that is equipped with a cooling system can range from $1,500 to $2,700, depending on the type of system being installed. In spite of the high initial costs of the ground-water heat pump, however, its lower operating cost makes it extremely cost-effective. Table 2.2 compares the operating costs of a ground-water heat pump with those of more conventional heating methods, including the air-source heat pump. It can be seen that the decreased operating costs of the ground-water heat pump can result in an accelerated payback schedule that can surpass those of other types of heating and cooling systems.

Closed-loop Heat Pumps

The closed-loop heat pump also utilizes water, drawn from a closed loop of piping that is filled with water and then buried underground. These types of heat pumps are often referred to as *earth-coupled* be-

----

[4]*Understanding Heat Pumps*, p. 22.

TABLE 2.2   Operating Costs of Heating Systems

Type of Energy	Units	Gross Heat Content	Usable Heat Content	Fuel Cost	Cost—10,000 Btu
#2 Fuel oil	gallons	140,000 Btu/gal.	84,000 Btu/gal.*	1.20/gal.	14.29¢
Propane gas	gallons	91,000 Btu/gal.	54,600 Btu/gal.*	71.38¢/gal.	13.07¢
Natural gas	therms	95,000 Btu/th	57,000 Btu/th*	35¢/th	6.14¢
Electricity (resistance)	kwh	3,413 Btu/kwh	3,413 Btu/kwh	4.192¢/kwh	12.28¢
Electricity (A/A heat pump standard model)	kwh	6,826 Btu/kwh	6,826 Btu/kwh**	4.192¢/kwh	6.14¢
Electricity (A/A heat pump efficiency)	kwh	8,533 Btu/kwh	8,533 Btu/kwh***	4.192¢/kwh	4.91¢
Electricity (W/A heat pump)	kwh	10.500 Btu/kwh	10,500 Btu/kwh****	4.192¢/kwh	3.99¢
Coal (Pocahontas)	ton	12,000 Btu/lb	5,000 Btu/lb*****	$125/ton	12.50¢

*Based on 60 percent efficiency
**Based on COP of 2 (200 percent efficiency)
***Based on COP of 2.5 (250 percent efficiency)
****Based on COP of 3.11 (311 percent efficiency)
*****Based on 38 percent efficiency

A/A = Air-to-air heat pump
W/A = Water-to-air heat pump

(Courtesy National Well Water Association)

cause they depend upon the exchange of heat between the earth and the water; the buried loop itself is often referred to as a *heat exchanger* for the same reason.

A great deal of research has been performed on closed-loop, earth-coupled heat pumps at Oklahoma State University, all of it spearheaded by James E. Bose, Director of Engineering Technology. Since the mid-70s, Bose has worked closely with The Charles Machine Works, a manufacturer of Ditch Witch trenchers and drills, to develop cost-effective methods of installing closed-loop systems. Their efforts have also resulted in standardized design calculations that enable an installer to determine the proper type of loop, the size and amount of piping needed, and the most efficient installation method, all based primarily on climate and soil characteristics in the area where the system is to be installed. Oklahoma State University has developed computer models for many U.S. cities to aid installers in system design and installation, and in determining the best type of pipe to use in closed-loop systems.

Closed-loop systems utilize water circulating through high-strength plastic pipe buried in the earth. The 50-year minimum life of the pipe with a safety factor of over 10 results in a buried pipe installation with a very long life. This is an improvement in many cases over those systems that use water from a well as the carrier of the earth's low-temperature heat to the heat pump. These are very effective systems as long as enough water flow is available, it can be returned to the aquifer without pollution problems, the water doesn't cause mineral deposits or corrosion, and it can be pumped from a shallow enough depth so that the pumping energy doesn't cancel out too much of the heat-pump energy savings. Water-well systems must be designed and installed properly to function dependably, minimize maintenance, and prevent pollution.

These problems are eliminated with a closed-loop system, which recirculates 50 to 150 gallons of water through sealed loops of pipe. The pipe is either buried 4 to 6 feet deep in trenches or is inserted with very closely fused U-bends into holes drilled into the earth 100 to 300 feet deep, where the 2,000 to 3,000 square feet of clear area is not available for a horizontal system. Horizontal systems have had the greatest application because they are easier to install.

The piping is placed in trenches dug by chain diggers and may consist of single or multiple pipes (Figure 2.8). In northern climates, single pipes are buried at depths such that the summer sun can melt any frozen soil caused by heat removal. In southern climates, several pipes

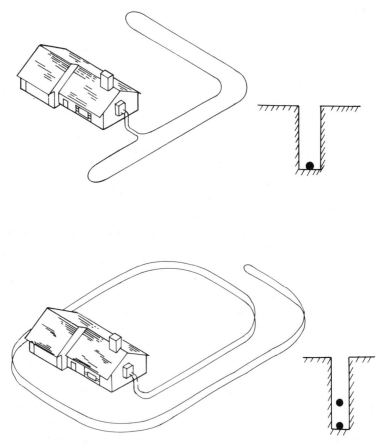

Figure 2.8   Installation of single and multiple piping (Courtesy National Well Water Association).

may be buried in a single trench to limit overall trench length, which may reduce overall cost as well. The allowable reduction in length will depend on soil conditions and seasonal operating times. Since trenches only 6 inches wide are used for horizontal installations, the environmental impact is negligible. The amount of heat removed from or dissipated into the earth by these systems represents a very small percentage of the heat capacity of the affected earth. Temperature sensors buried at various distances from the pipes typically show soil temperature changes compared to the normal temperature of the soil at that depth of less than $10°F$ at the pipe surface to $1°$ or $2°$ at 5 feet from the pipe. The thermal impact on the earth is much less than that of building foundations, sidewalks, driveways, parking lots and streets, which conduct huge amounts of heat into and out of the earth.

The development of vertical loop systems has significantly increased the number of potential users of closed loops due to the reduced land surface area required. However, the vertical-loop system (Figure 2.9) requires special consideration to protect the environment. This involves using a cement or bentonite clay plug in the top of the hole to prevent surface water intrusion into a ground water aquifer, and the injection of a cement or bentonite slurry into the bore hole to prevent water from one ground water aquifer from flowing to another one at a different

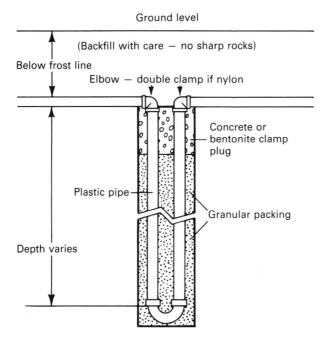

Figure 2.9 Special installation methods to protect the environment (Courtesy National Well Water Association).

depth, possibly causing pollution of the aquifers. This is accomplished by pumping the slurry to the bottom of the hole through a plastic pipe previously inserted with the loop, with its end taped near the bottom of the loop. After the hole has been filled from the bottom by pumping the slurry through the pipe, the pipe is pulled loose, flushed with water, and coiled for reuse.

In other areas of the country, it is necessary for an antifreeze solution to be added to the water in the closed loop to prevent freezing of the circulating water and to allow the system to gain capacity and efficiency by using the large amount of heat released when the water is frozen. There have been concerns expressed about possible leakage. The 20% solution of nontoxic calcium chloride or propylene glycol in the small 50 to 150 gallon quantities typically used for residential systems presents a minimal potential for objectional impact on the earth, even if it were to leak out. The possibility of a leak in a buried-pipe system using the recommended grade of pipe, type of fused joints, and testing procedure is slight. The pipe used has a minimum design life of 50 years, and the fused joints that are used are actually stronger than the pipe itself. The high density polyethylene pipe and the fusion joint systems used in closed loops is the same system that has been used for natural gas distribution for 30 years without serious problems.

The installation of a closed-loop system will require the services of a qualified contractor who will be responsible for the exterior portion of the system. He will determine the type of loop based on land area and soil and climate conditions. Since closed-loop systems are relatively new, it is recommended that the installer contact either Oklahoma State University or The Charles Machine Works (see Appendix A) to obtain the names of contractors with an understanding of closed-loop systems. Both OSU and The Charles Machine Works offer intensive training programs in an effort to increase knowledge about closed-loop systems.

It should be understood that the type of loop configuration used will have no significant impact on operating efficiency, although each type has its own advantages and disadvantages. Installing horizontal loops is simpler, and requires lower-cost equipment and less training. However, they require longer lengths of pipe due to variations in soil temperature and moisture content; installations can be affected by extensive rainy weather. Also, a much larger area is required and hard rock will prevent their installation. Vertical loops require highly trained operators for the drilling machines, but less pipe length is required, which can offset much of the higher drilling cost. Also, a relatively smaller area is required.

As with the ground-water heat pump, the closed-loop heat pump, once installed, will result in significant reductions in operating costs when compared with an air-source heat pump, although initial costs are considerably higher. It is anticipated that these installation costs will be reduced as more systems are installed.

Three

Heat Pump Components and Circuitry

In this chapter, we will look at the components and circuitry that make up a heat pump. It should be understood that although many of the same components are used throughout the industry, some will differ depending on the manufacturer, as will their exact location within the heat pump. The installer should always refer to the manufacturer's installation manual, specification sheets, and component troubleshooting charts for the specifics of a particular heat pump.

3.1 THE COMPRESSOR

The compressor is often referred to as the heart of the heat pump, since it provides the primary means of controlling refrigerant flow throughout the system. A compressor is an electrically driven vapor pump that is used to circulate the refrigerant through the heat pump system. With the assistance of refrigerant control devices, it provides a pressure differential in the system when and where it is needed. Most compressors are hermetically sealed, meaning they are not serviceable in the field. In a hermetically sealed compressor, the motor is sealed in the same housing as the compressor, assuring no refrigerant leakage. (Motors will be discussed later in this chapter.)

Compressors can be classified as *reciprocating, rotary*, or *centrifugal*, with the majority of heat pumps today using either a reciprocating or rotary type of compressor. The design of the reciprocating compressor is somewhat similar to a modern automotive engine, with a piston driven from a crankshaft making alternate suction and compression strokes in a cylinder equipped with suction and discharge valves. Since the reciprocating compressor is a positive displacement pump, it is suitable for small displacement volumes, and is quite efficient at high condensing pressures and high compression ratios. Other advantages are its adaptability to a number of different refrigerants, the fact that liquid refrigerant may be easily run through connecting piping because of the high pressure created by the compressor, its durability, basic simplicity of design, and relatively low cost.

The rotary compressor (Figure 3.1) is used in some heat pumps today. This type of compressor operates through application of a circular motion rather than the reciprocating motion just discussed. Rotary compressors may be of fixed or rotating vane, as shown in Figure 3.1. Advantages are that a rotary compressor can usually be used to pump a deeper vacuum than a reciprocating compressor and, as a rule, is quieter during operation.

3.2 REFRIGERANT CONTROL DEVICES

Refrigerant is metered at the coil that is to absorb heat. During the heating cycle, the refrigerant must be metered at the outdoor coil, while during the cooling cycle, it must be metered at the indoor coil.

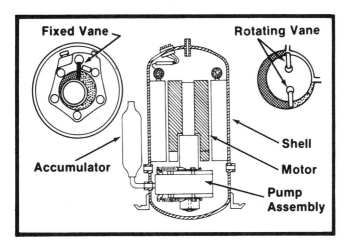

Figure 3.1   Rotary compressor.

There are many types of metering devices that may be used. In some cases, the refrigerant flow is controlled to each coil by a single metering device that allows flow in either direction. Such a device would be located in the liquid line. The direction that the refrigerant flows would reverse as the heat pump changes from heating to cooling, or vice versa. A common method of metering refrigerant to each coil is to use two separate metering devices and two check valves, one for each coil. The metering devices used could be expansion valves, capillary tubes, or a combination of the two.

Check valves (Figure 3.2) are used to prevent refrigerant from reversing its direction of flow during an off cycle, or during a change in the operating cycle. A simple spring-loaded valve (Figure 3.3) allows flow in one direction only and closes if pressures are such that reverse flow could occur. Check valves may be used in either liquid or gas lines and are frequently used to prevent backflow of liquid refrigerant or hot gas in low temperature controls.

The most commonly used device for controlling refrigerant flow is the *thermostatic expansion valve*. This device is able to control refrigerant flow by means of internal components that are responsive to the temperature and pressure of the refrigerant. Figure 3.4 shows the placement of a thermostatic expansion valve in a heat pump system.

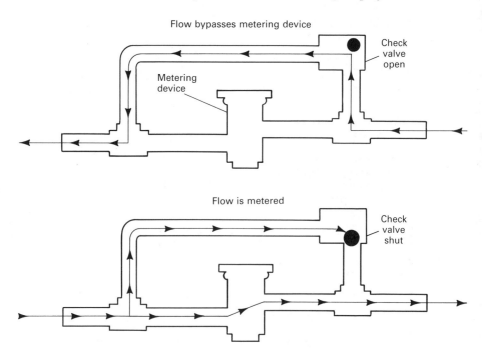

Figure 3.2   Check valves (Courtesy Carrier Corporation).

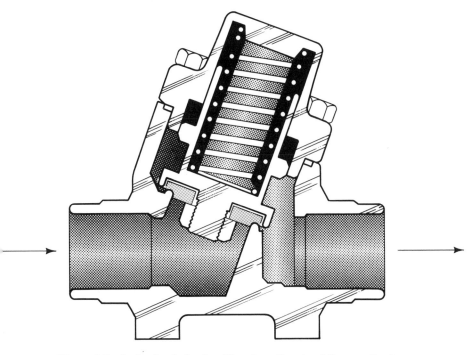

Figure 3.3   Spring-loaded valve (Courtesy Copeland Corporation).

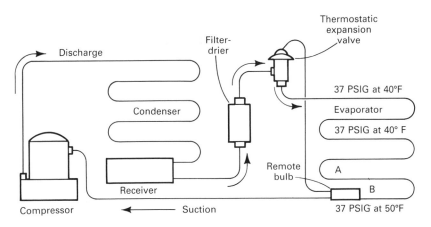

Figure 3.4   Thermostatic expansion valve (Courtesy Alco Controls Division, Emerson Electric Company).

Three forces govern an expansion valve's operation (Figure 3.5). They are: the pressure created by the remote bulb and power assembly (P_1), the coil pressure (P_2), and the equivalent pressure of the superheat spring (P_3). The remote bulb and power assembly is a closed system,

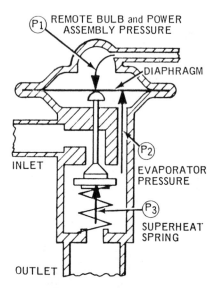

REMOTE BULB and POWER ASSEMBLY PRESSURE (P_1)

DIAPHRAGM

INLET

(P_2) EVAPORATOR PRESSURE

(P_3)

SUPERHEAT SPRING

OUTLET

Figure 3.5 Detail of thermostatic expansion valve (Courtesy Alco Chemical Division, Emerson Electric Company).

and for purposes of this discussion, we will assume that the remote bulb and power assembly charge is the same refrigerant as that used in the system. We will also assume that the heat pump is in the cooling cycle and that the indoor coil is acting as the evaporator coil.

The pressure within the remote bulb and power assembly (P_1) then corresponds to the saturation pressure of the refrigerant temperature leaving the coil acting as the evaporator and moves the valve pin in the opening direction. Opposed to this force on the underneath side of the diaphragm and acting in the closing direction is the force exerted by the evaporator pressure (P_2) and the pressure exerted by the superheat spring (P_3). The valve will assume a stable control position when these three forces are in equilibrium $(P_1 = P_2 + P_3)$. As the temperature of the refrigerant at the evaporator outlet increases above the saturation temperature corresponding to the evaporator pressure, it becomes superheated. The pressure thus generated in the remote bulb and power assembly increases above the combined pressures of the evaporator pressure and the superheat spring, causing the valve pin to move in the opening direction. Conversely, as the temperature of the refrigerant leaving the evaporator coil decreases, the pressure in the remote bulb and power assembly also decreases, and the combined evaporator and spring pressures cause the valve pin to move in the closing direction.

The factory superheat setting of a thermostatic expansion valve is made with the valve pin just starting to move away from the seat. The valves are designed so that an increase in superheat of the refrigerant leaving the evaporator coil is necessary for the valve pin to open to its rated position. For example, if the factory setting is 10°F superheat, the operating superheat at the rated open position (full load rating of the valve) will be $14°F$ superheat. If the system is operating at half load, with 50% compressor capacity reduction, the valve will operate at $12°F$ superheat.

As the operating superheat is raised, the evaporator coil's capacity decreases, because more of the coil surface is required to produce the superheat necessary to open the valve. It is most important to adjust the operating superheat correctly, and that a minimum change in superheat is required to move the valve pin to full open. Accurate and sensitive control of the liquid refrigerant flow to the coil acting as the evaporator is necessary to provide maximum capacity under all load conditions.

Capillary tubes may also be used for refrigerant control. A capillary tube is a length of tubing of small diameter, with the internal diameter held to extremely close tolerances. It is used as a fixed orifice to perform the same function as an expansion valve; that is, it separates the high and low pressure sides of the system and meters the proper feed of refrigerant.

Since the orifice is fixed, the rate of feed is relatively inflexible. Under conditions of constant load and constant discharge and suction pressures, the capillary tube performs very satisfactorily. However, changes in load or fluctuations in pressure can result in overfeeding or underfeeding.

A major advantage of the capillary tube in some systems is the fact that refrigerant continues to flow into the coil even after the compressor stops, thus equalizing pressures on the high and low sides of the system. This allows the use of low-starting torque motors.

The refrigerant charge is critical in capillary tube systems. Too much refrigerant will cause high discharge pressures and motor overloading and possible liquid floodback to the compressor. Too little refrigerant will allow vapor to enter the capillary tube, causing a loss in system capacity. However, due to its basic simplicity, a capillary tube system is considered to be the least expensive type of refrigerant control system.

Compressor Service Valves

Compression suction and discharge service valves are shut-off valves with a manual operated stem. Most service valves are equipped with a gauge port so that refrigerant operating pressure may be observed.

When the valve is back-seated (the stem turned all the way out), the gauge port is closed and the valve is open (See Figure 3.6). If the valve is front-seated (the stem turned all the way in), the gauge port is open to the compressor and the line connection is closed. In order to read the pressure while a compressor is operating, the valve should be back-seated and then turned in one or two turns in order to slightly open the connection to the gauge port. The compressor is always open

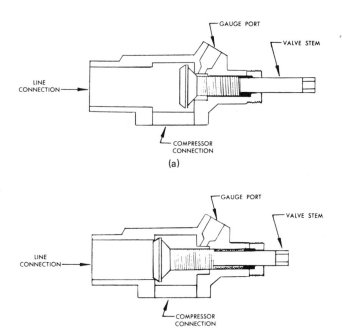

Figure 3.6 Compressor service valves back-seated (a) and front-seated
(b) (Courtesy Copeland Corporation).

to either the line or the gauge port, or both if the valve is neither front-
nor back-seated.

Schrader Type Valves

Schrader type valves (Figure 3.7) are used for convenient checking of
system pressure where it is not economical, convenient, or possible to
use the compressor service valves with gauge ports. The Schrader type

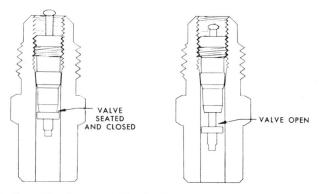

Figure 3.7 Schrader type valve in closed position (left) and open position (right) (Courtesy Copeland Corporation).

valve is similar in appearance and principle to the air valve used on auto-
mobile or bicycle tires, and must have a cap for the fitting to insure
leak-proof operation. This type of valve enables checking of the system
pressure or charging refrigerant without disturbing the unit operation.
An adapter is necessary for the standard serviceman's gauge or hose
connection to fit the Schrader type valve.

Pressure Relief Valves

Safety relief valves are required by many local construction codes. Various types of relief valves are available, and the system requirement may
be dictated by the local code requirement. Normally, code requirements specify that the ultimate strength of the high side parts shall be a
minimum of 5 times the discharge or rupture pressure of the relief valve,
and that all condensing units with pressure vessels exceeding 3 cubic
feet interval volume shall be protected by a pressure relief device. Discharge may be to the atmosphere, or it may be a discharge from the
high-pressure side of the system to the low-pressure side.

Water Regulating Valves

In water-source heat pump systems, a modulating water regulating valve
(See Figure 3.8) is normally used to economize on water usage and control pressures within reasonable limits. Water valves may be either pressure or temperature actuated.

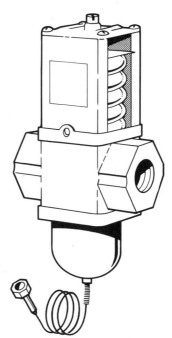

Figure 3.8 Water regulating valve
(Courtesy Copeland Corporation).

The Reversing Valve

In order to make the process of transferring heat work for both the heating and cooling cycles, the direction of refrigerant flow must be controlled. The device that controls the direction of refrigerant flow in a heat pump is called the *reversing valve*. A reversing valve is one type of solenoid valve and is often called a *four-way valve*, because it has four pipe connections. Before explaining the operation of the reversing type of solenoid valve, let's first look at solenoid valves in general.

Referring to Figure 3.9, a solenoid valve consists of two distinct but integral acting parts, a solenoid and a valve. The solenoid is nothing more than electrical wire wound in a spiral around the surface of a cylindrical form, usually of circular cross-section. When an electrical current is sent through the windings, they act as an electromagnet. The force field that is created in the center of the solenoid is the driving force for opening the valve. Inside is a moveable magnetic steel plunger that is drawn toward the coil when energized. The valve contains an orifice through which fluid flows when open. A needle or rod is seated on or in the orifice and is attached directly to the lower part of the plunger.

When the coil is energized, the plunger is forced toward the center of the coil, thus lifting the needle valve off of the orifice and allowing flow. When the coil is de-energized, the weight of the plunger and, in some designs, a spring causes it to fall and close off the orifice, thus stopping the flow through the valve. Figure 3.10 shows a simple schematic of a typical solenoid valve in operation.

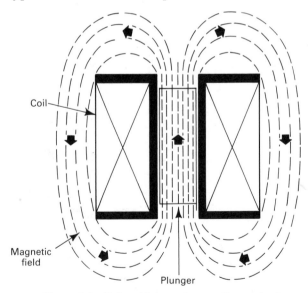

Figure 3.9 Solenoid operation: coil energized.

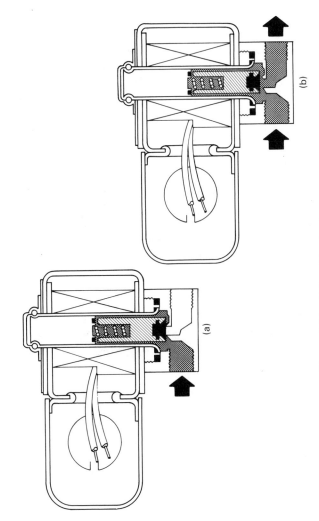

Figure 3.10 Solenoid valve in operation.

The reversing valve has four pipe connections (see Figure 3.11). Flow of refrigerant through two of these connections never changes; these connections are the hot gas discharge from the compressor and the suction line back to the compressor. The remaining two ports connect to the outdoor and indoor coils. The direction of refrigerant flow depends on the cycle. The valve is solenoid-controlled.

A typical reversing valve directs the refrigerant flow by using a free-floating slide inside a cylinder. The slide changes position by refrigerant pressure. As this slide shifts, it directs the refrigerant flow to and from the coils (see Figure 3.12). The compressor discharge is connected to the single port on the cylinder. There is always discharge pressure bleeding between the two ends of the slide, represented in Figure 3.12 by arrows. At each end of the slide, there is a small orifice that allows the discharge gas to bleed behind the slide.

At both ends of the cylinder, there are capillary lines that connect to a pilot solenoid chamber. Also connected to the pilot solenoid chamber is another capillary that connects to the compressor suction line. In Figure 3.13, it's the center capillary. The pilot solenoid pin carrier changes position when the solenoid coil is energized or de-energized.

The typical reversing valve may be piped into the refrigerant system to provide either heating or cooling when the pilot solenoid coil is

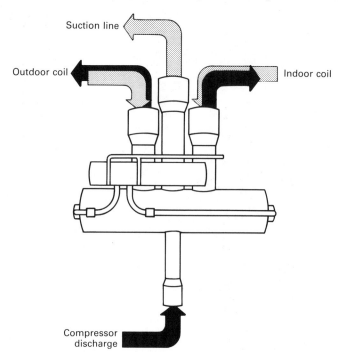

Figure 3.11 The four connections of the reversing valve (Courtesy Carrier Corporation).

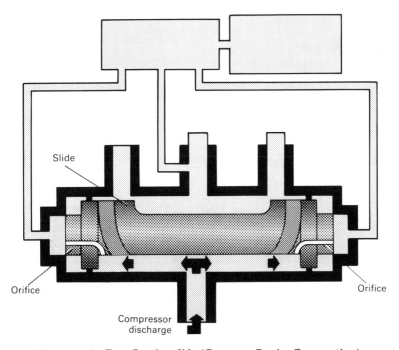

Figure 3.12   Free-floating slide (Courtesy Carrier Corporation).

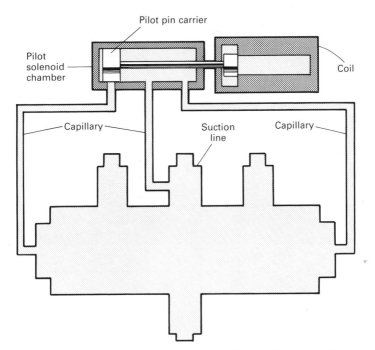

Figure 3.13   Capillary lines (Courtesy Carrier Corporation).

energized electrically. For our example, we will show the reversing valve pilot solenoid energized electrically during the cooling cycle, and de-energized during the heating cycle. Referring to Figure 3.14, starting at (1), the solenoid is de-energized, so the pilot solenoid pin carrier moves to the left. At (2), the compressor discharge gas bleeds behind the left-hand side of the slide. At (3), the discharge gas is trapped in the capillary tube. The pressure builds until it reaches the discharge pressure.

Referring to Figure 3.15, at (4), the pressure behind the right side of the slide equals the compressor suction pressure, which passes through the pilot solenoid chamber and down through the center capillary to the suction line at (5).

Since the pressure behind the right side of the slide is less than the pressure behind the left side, the slide moves to the right (see Figure 3.16). The hot compressor discharge gas is directed to the indoor coil and the cold suction gas is directed from the outdoor coil to the compressor. The heat pump is therefore in the heating cycle.

Referring to Figure 3.17, when energizing the solenoid valve, at (1), the pilot pin carrier moves to the right. Compressor suction pressure now passes through the center capillary and the pilot chamber down to the left side of the slide at (2). At (3), compressor discharge pressure bleeds through the orifice behind the right side of the slide,

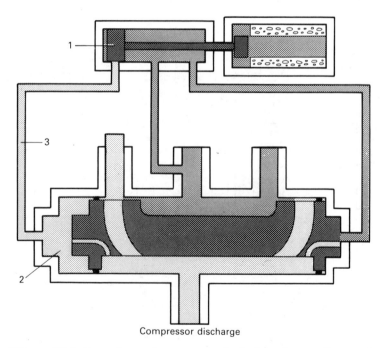

Compressor discharge

Figure 3.14 Reversing valve energized during the cooling cycle (Courtesy Carrier Corporation).

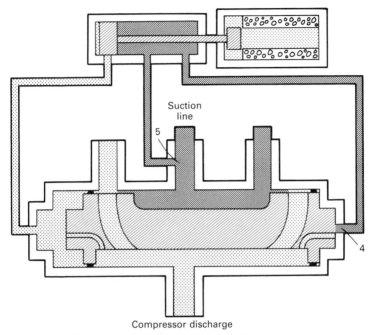

Figure 3.15 Pressure behind the right side of the slide equals compressor suction pressure (Courtesy Carrier Corporation).

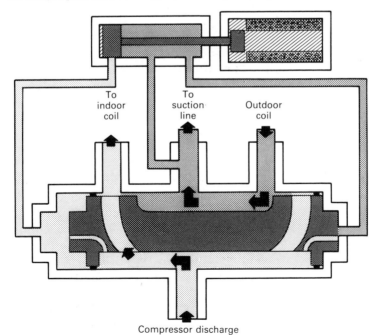

Figure 3.16 Pressure behind the right side is less than pressure behind the left side (Courtesy Carrier Corporation).

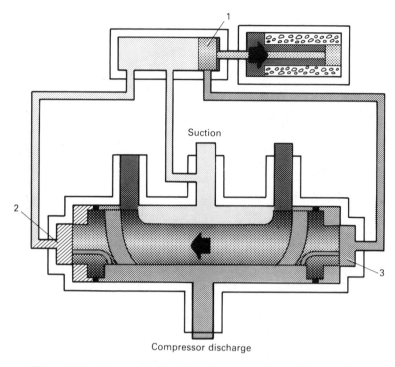

Figure 3.17   Energizing the valve (Courtesy Carrier Corporation).

where it is trapped. The pressure behind the left side of the slide is now less than the pressure behind the right side, and the slide moves to the left. The reversing valve is now directing the hot compressor discharge gas to the outdoor coil and the cold suction gas from the indoor coil to the suction accumulator (see Figure 3.18). The heat pump therefore is in the cooling cycle.

The Accumulator

Some heat pumps incorporate an *accumulator* (Figure 3.19) located in the suction line between the compressor and the reversing valve. In this position, the accumulator provides a surge chamber and a reservoir to contain and control liquid refrigerant and oil that would otherwise enter the compressor directly. The accumulator consists of a screen, a screen retainer, and a stand pipe. An oil bleed hole is located at the base of the stand pipe to allow a controlled amount of oil to return to the compressor.

The accumulator becomes most important during the heating cycle when the outdoor coil is being used as an evaporator. At the colder temperatures, the outdoor coil may not be able to evaporate all the

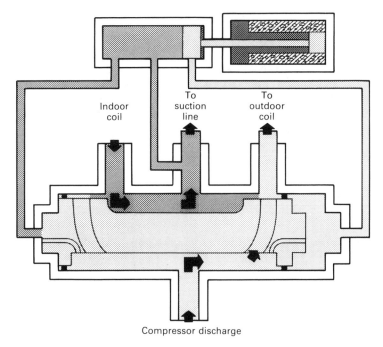

Figure 3.18 Reversing valve (cooling cycle). (Courtesy Carrier Corporation).

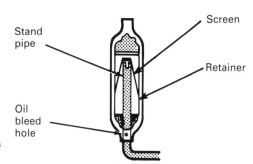

Figure 3.19 Internal components of accumulator.

refrigerant and will be flooded. This situation may also occur at the end of a defrost cycle in the case of an air-source heat pump. Any liquid carryover will be caught in the accumulator.

Dehydrators

Moisture is one of the basic enemies of a heat pump system, and the moisture level must be held to an acceptable low level to avoid system malfunctions or compressor damage. Even with the best precautions, moisture will enter a system any time it is opened for field service. Unless the system is thoroughly evacuated and recharged after exposure to moisture, the only effective means of removing small amounts of moisture is with a dehydrator.

Dehydrators, or driers as they are commonly called, consist of a shell filled with a dessicant or drying agent, with an adequate filter at each end. Some driers are made in porous block form so that the refrigerant is filtered by the entire block. Driers are mounted in the refrigerant liquid line (Figure 3.20), so that all of the refrigerant in circulation must pass through the drier each time it circulates through the system. Most driers are constructed so that they can serve a dual function as both filter and drier.

Many different drying agents are used, but practically all modern driers are either of the throw-away or of the replaceable-element type. It is considered good practice to discard the used drier element each time the system is opened and replace it with a new drier or drier element. Figure 3.21 shows a typical filter-drier.

The Crankcase Heater

When a heat pump is in the off cycle, refrigerant will migrate to the oil in the compressor crankcase. This migration is faster during the heating cycle when the compressor is coldest. The liquid refrigerant could load up in the compressor crankcase. When it tries to start again, this could

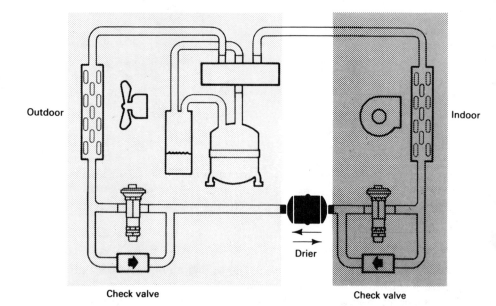

Figure 3.20   Location of a drier.

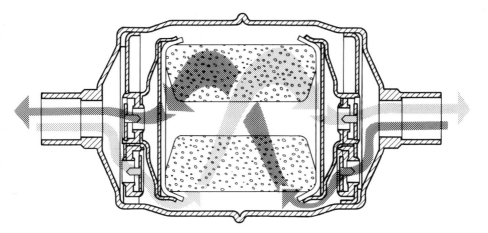

Figure 3.21   Filter-drier.

cause a hard starting problem and/or physical damage due to loss of lubrication and slugging.

To prevent this, heat is applied to the crankcase area either by means of a trickle circuit through the run winding of the compressor motor or by a crankcase heater located on the compressor. These heaters can be immersed into the compressor oil or mounted externally. Depending on the compressor size, they will vary from a few watts to several hundred watts. The heater will raise the temperature of the oil so that the absorption of refrigerant into the oil on shutdown is kept at a minimum. They may be on all the time or only when the compressor is not running, depending on design.

Discharge Mufflers

On systems where noise transmission must be reduced to a minimum or where compressor pulsation might create vibration problems, a discharge muffler is frequently used to dampen and reduce compressor discharge noise. The muffler is basically a shell with baffle plates, with the required internal volume dependent primarily on the compressor displacement, although the frequency and intensity of soundwaves are also factors in muffler design.

Evaporator/Condenser Coils

In an air conditioning unit, the evaporator and condenser coils, although similar in construction, are designed for different functions. The evaporator coil is used to evaporate liquid refrigerant with a low internal pressure drop, while the condenser coil is used to condense vapor refrigerant with a comparatively high pressure drop. During the heating cycle of a heat pump's operation, however, the coil functions are reversed; i.e., the evaporator becomes the condenser and the condenser becomes the evaporator. Due to this reversal, the coils are specifically engineered for both condensing and evaporating, and the two coils are normally referred to as the indoor coil and the outdoor coil to avoid confusion. These coils are also commonly referred to as *heat exchangers.*

The indoor coils in both air- and water-source heat pumps are very similar. Most are constructed of copper or aluminum tubing (through which the refrigerant flows) that is equipped with closely spaced fins (Figure 3.22). This type of construction is widely accepted, with variations depending on the manufacturer, because it has proven to be the most efficient means of providing heat exchange.

The outdoor coil of an air-source heat pump is similar in design and construction to the indoor heat exchanger. However, in a water-source heat pump, a different type of construction is used (Figure 3.23). Here, because water replaces air as the heat source, a circular configura-

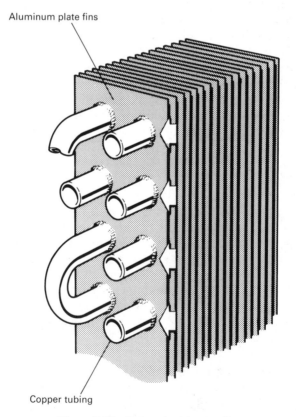

Aluminum plate fins

Copper tubing

Figure 3.22   Indoor/outdoor coil.

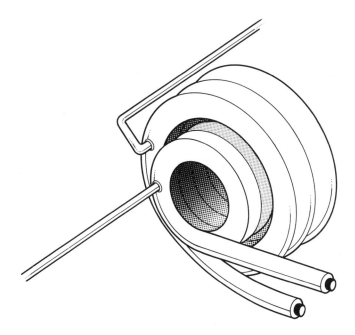

Figure 3.23 Water coil.

tion is used. This type of configuration allows the water and refrigerant, traveling in opposite directions within the coil, to exchange heat; this configuration is used in all water-source heat pumps today.

3.3 ELECTRICAL CONTROL DEVICES

The basic function of most electrical control devices is to make or break an electric circuit, which in turn controls a contactor, a solenoid coil, or some other functioning part of the system. Controls are available that can make or break a circuit on either a rise or fall in pressure or temperature. The type of action required depends on the function of the control and the medium being controlled.

The point at which a control closes a contact and makes a circuit is called the *cut-in point*. The point at which the control opens the switch and breaks the circuit is called the *cut-out point*. The difference between the cut-in and cut-out points is known as the *differential*.

A very small differential maintains close control but can cause short cycling of the compressor. A large differential will give a longer running cycle, but may result in fluctuations in the pressure or temperature being controlled. The final operating differential must be a compromise.

The differential may be either fixed or adjustable, depending on the construction of the control. Adjustment of controls varies, depending on type and manufacturer. On some controls, both the cut-in and cut-out points may be set at the desired points. On many pressure controls, the differential can be adjusted, and this in turn may affect either the cut-in or the cut-out point.

Line Voltage and Low Voltage Controls

Line voltage controls are designed to operate on the same voltage as that supplied to the compressor. Both 110 and 220 volt controls are commonly used. Local codes often require low voltage controls, and a control circuit transformer may be used to reduce line voltage to the control circuit voltage.

Low Pressure and High Pressure Controls

A low pressure control is actuated by the refrigerant suction pressure and normally is used to cycle the compressor for capacity control purposes, or as a low limit control. The low pressure control often is used as the only control on small systems that can tolerate some fluctuations in the temperature to be maintained. The standard low pressure control makes a circuit on a rise in pressure and breaks on a fall in pressure.

Thermostats

A thermostat acts to make or break a circuit in response to a change in temperature. There are numerous types of thermostats available for use with heat pumps, and most manufacturers offer a number of thermostats that are compatible with their equipment.

In a heat pump system, there may be more than one temperature-control device. All heat pumps will incorporate a standard heating/cooling thermostat (Figure 3.24) that is installed indoors. This thermostat is set to the desired indoor temperature, and heating or cooling takes place as needed.

Heat pumps that utilize electric resistance heat as the supplemental heat source may incorporate a separate thermostat (Figure 3.25) that controls the operation of the resistance heaters. This thermostat, which is installed outdoors, is set based upon calculations that determine the temperature at which supplemental heat is required. Some standard heat pump thermostats are equipped with one or possibly two manually operated switches (see Figure 3.23) designed into the system to control supplemental heat. Use of either of these switches interrupts the compressor and allows the supplemental heating system to supply heat, although the system will still be under temperature control.

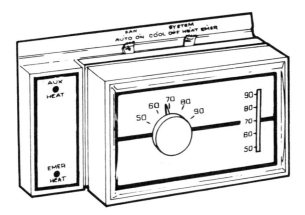

Figure 3.24   Heat pump indoor thermostat.

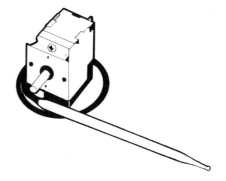

Figure 3.25   Resistance heat thermostat.

Air-source heat pumps that require defrost circuitry may also incorporate a temperature-activated control (Figure 3.26) that provides for automatic de-icing of the outdoor coil. This is one method of defrost control; other methods will be discussed later in this chapter.

Figure 3.26   Temperature-activated control.

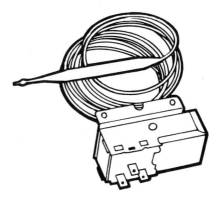

Electric Resistance Heaters

Like any refrigeration system, a heat pump will start to lose capacity when the suction pressure drops. With a heat pump operating in the heating cycle, this happens as the outdoor temperature drops. This also causes the efficiency and COP to drop. The reason for this is that the gas density of the refrigerant decreases as the suction pressure decreases. This means that the refrigerant weighs less per cubic foot, and so the compressor pumps less pounds of refrigerant through the system.

Each pound of refrigerant is capable of absorbing a certain quantity of heat. With less pounds of refrigerant circulated, the capacity would therefore decrease. Along with a decrease in the gas density at lower suction pressures, the re-expansion stroke of the compressor piston is greater. This results in a shorter effective stroke and thus, a decrease in flow rate.

However, as the outdoor temperature starts to drop, so does the capacity of the system. On colder days when the capacity drops below the required heat loss of the conditioned space, additional heat must be supplied. The most common way to provide supplemental heat is to use electric resistance heaters (Figure 3.27).

The capacity of electric resistance heaters will depend on the size of the heat pump and the area where it is installed. When air-source heat pumps are installed in colder climates, two or three heaters may be necessary not only to provide additional heat, but also to temper the discharge air during the defrost cycle. In warm weather climates, supplemental heat may not be required at all.

In colder climate areas, the electric utility or a local code may require that electric heat be installed with enough capacity to provide comfort in an emergency situation, such as if the heat pump (air or water source) fails to operate. The emergency heat capacity requirement may vary from 60–100% of the heat required at design conditions.

An outdoor thermostat is usually used to control the operation of

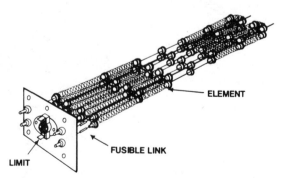

Figure 3.27   Electric resistance heater element.

the heaters, although some heaters are equipped with their own controls. These thermostats are usually adjustable. When more than one thermostat is used, staging of strip heat capacity to meet the load requirements can be provided. Setting of the outdoor thermostats is determined by the building construction and outdoor air temperature.

A good heating load estimate will indicate the balance point. The *balance point* is the point at which the unit capacity is equal to the heat loss of the building. At any outdoor temperature below the balance point, auxiliary heat will be necessary to assist the heat pump in maintaining comfort. The outdoor thermostats are usually set about $2°F$ above the balance point.

Let's take an example of a typical 3-ton air-source heat pump. The heat pump capacity and the outdoor temperature have been plotted in Figure 3.28. Note that at $50°F$ outside, the heat pump has a capacity of approximately 45,000 BTU/hr. As the outdoor temperature drops to $10°F$, the unit capacity drops to approximately 21,000 BTU/hr.

In Figure 3.29, the building heat loss is plotted. Note that at about $65°F$, the heat loss or capacity required is zero. This is the point where neither cooling nor heating is required. The point at which the unit capacity line and the building heat-loss line intersect is the balance point. In this example, the balance point is at approximately $32.5°F$ outdoor temperature, and the heat loss is equal to the unit capacity of approximately 34,500 BTU/hr.

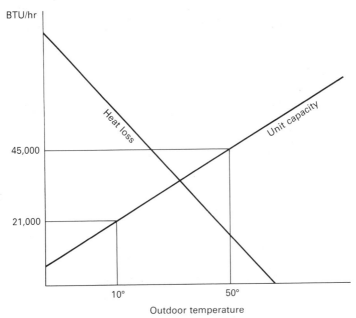

Figure 3.28   Capacity versus heat loss (Courtesy Carrier Corporation).

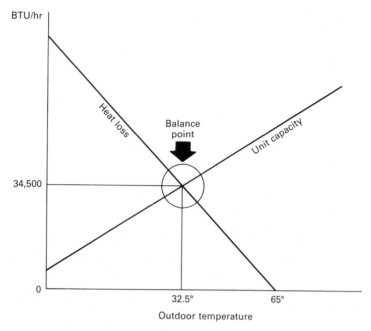

Figure 3.29 The balance point (Courtesy Carrier Corporation).

In this example, let's pick a design point of $10°F$ outdoor temperature. Referring to Figure 3.30, at $10°F$, the heat pump will require an additional 36,500 BTU/hr to maintain comfort. The heater selected must deliver at least 36,500 BTU/hr at the design point.

A heater must also be selected on the basis of emergency heat in the event of a heat pump failure. In this example, let's pick a heater with 80% emergency heat backup, although local codes will determine the percentage of backup heat that is required. Referring to Figure 3.31, at the design point of $10°F$, the capacity required would be approximately 57,500 BTU/hr. This would mean an electric heater capacity of 46,000 BTU/hr, which works out to be slightly under 13.5 kilowatts.

Heat pump manufacturers offer electric resistance heating element kits as optional accessories for their equipment. The 3-ton heat pump we have selected in this example has two heater packages available, using 9 kilowatts and 13.5 kilowatts of power, respectively. Obviously, the 9 kilowatt package would not supply enough heat to satisfy the 80% emergency heat requirement. The 13.5 kilowatt heater has three stages of capacity of 4.5 kilowatts each, or 15,360 BTU/hr per stage, as shown in Figure 3.32. These capacities are rated at 240 volts, and correction factors are available for lower voltages.

Let's energize the first stage of electric heat at our balance point.

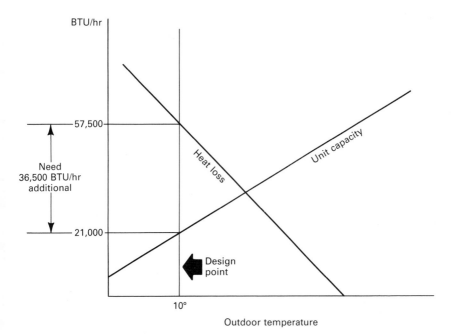

Figure 3.30 Additional BTUs necessary to maintain comfort (Courtesy Carrier Corporation).

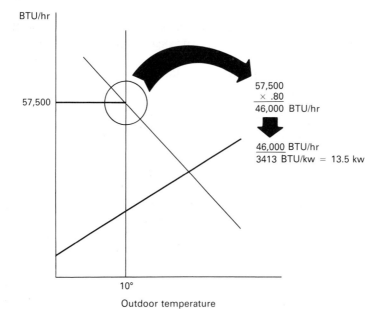

Figure 3.31 Amount of auxiliary heat needed in the example (Courtesy Carrier Corporation).

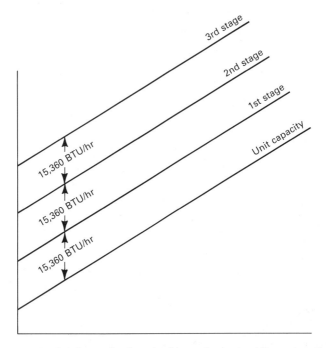

Figure 3.32 Staging of electric heat elements (Courtesy Carrier Corporation).

Referring to Figure 3.33, this will give us an additional 15,360 BTU/hr capacity and a new balance point. The heat pump capacity plus the first stage of electric heat will maintain the condition down to 22.5°F. At any temperature below 22.5°F, additional heat will be required. At this new balance point, let's energize the second stage of strip heat, or add another 15,360 BTU/hr.

Referring to Figure 3.34, we now have a new balance point at 13°F. The heat pump plus the first and second stages of electric heat will maintain conditions down to this temperature. At any temperature below this balance point, additional heat will be necessary.

Let's now energize the third stage. As shown in Figure 3.35, with the heat pump and all three stages of electric heat operating, the new balance point will now maintain conditions down to 4°F, which is well below the design point. We have now met the design point, and have enough electric heat capacity to meet the 80% emergency heat required.

As mentioned earlier, some electric heater kits are equipped with their own controls, eliminating the need for outdoor thermostats. Some newer models offer remote-control panels to allow for remote operation as well. Electric resistance heat kits are normally installed near the indoor blower of the heat pump and can be "slid" into place quite easily.

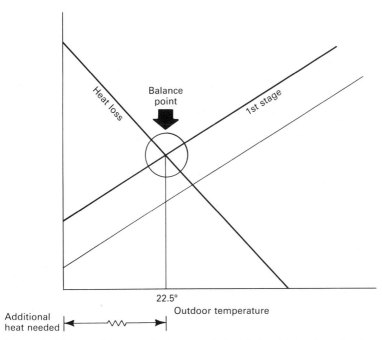

Figure 3.33 Energizing the first stage of electric heat (Courtesy Carrier Corporation).

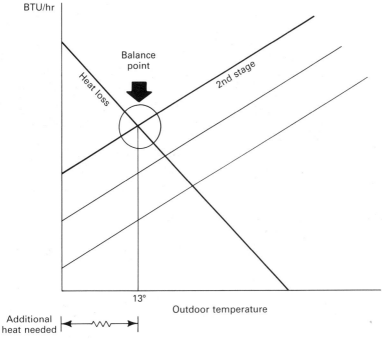

Figure 3.34 New balance point (Courtesy Carrier Corporation).

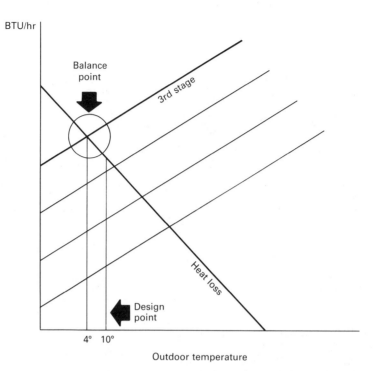

Figure 3.35 Energizing the third stage of electric heat (Courtesy Carrier Corporation).

Defrost Cycle Initiation and Termination

There are several methods currently used for defrost cycle initiation and termination in air-source heat pumps. It is important to understand thoroughly the method used in the heat pump being installed or serviced.

One method incorporates a temperature sensor (as discussed earlier in this chapter) that "senses" the accumulation of frost on the outdoor coil at temperatures below 32°F. Another method senses the air pressure difference between the outside and inside surfaces of the outdoor coil. Frost on the outdoor coil reduces the space between the fins, thus causing a restriction of the flow of air over the coil. Some recently developed sensing methods incorporate photoelectric sensing "eyes" and electric sensors that "see the frost" or measure the electrical conductivity that is passed through the frost accumulation. These sensors feed information to either a "time-and-temperature" or "demand" defrost system.

A demand defrost system defrosts the heat pump only when enough frost is present to significantly decrease heat pump efficiency.

This could range from every 30 minutes during a driving snowfall to every 7 or 8 hours in bitter cold, low-humidity conditions.

A time/temperature defrost system searches for frost buildup at a given time interval (usually 90 minutes). If the defrost thermostat senses temperatures below freezing, the defrost timer will initiate the defrost cycle during that search period. If above freezing temperature is sensed, the timer will wait another time interval (usually 90 minutes) to make another search for frost accumulation.

Initiating defrost energizes a special, multicontact relay. This relay ensures that the reversing valve is in the cooling mode, thus diverting hot compressor discharge gas to the outdoor coil. Simultaneously, the outdoor fan motor is turned off, facilitating the frost removal. Because the indoor coil is functioning as the evaporator during defrost, the defrost relay energizes one or more banks of supplemental heat to overcome the cold air leaving the indoor coil (see Figure 3.36).

Defrost termination is usually accomplished by the same frost-sensing devices described earlier. The sensing devices will terminate the defrost cycle when the outdoor coil has sufficiently cleared itself of frost. Most defrost control systems have a 10 minute override feature that ensures positive defrost termination after 10 minutes if the frost sensing device has not already done so. This condition might exist at extremely low ambient temperatures, during steady, strong prevailing winds, or with low refrigerant charge. Upon defrost cycle termination, the defrost controls place the heat pump back into the heating mode; i.e., the reversing valve shifts to the heating mode, the outdoor fan is energized, and any unnecessary strip heaters are de-energized.

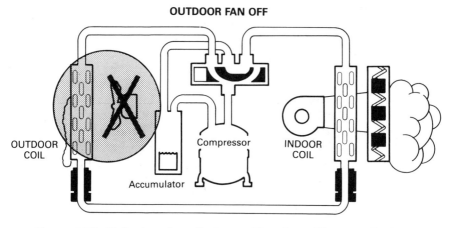

Figure 3.36   Defrost cycle activates auxiliary heat (Courtesy Carrier Corporation).

3.4 MOTORS AND STARTING EQUIPMENT

Electric motors are used as the power source for the compressor in a heat pump and also for the fan/blower assemblies. In an air-source heat pump, there are normally three electric motors: one in the compressor compartment, one to power the outdoor fan (which circulates air over the outdoor coil), and one to power the indoor fan/blower assembly (which delivers warm or cool air to the home). In a water-source heat pump, there will normally be only two electric motors: one in the compressor compartment and one to power the indoor fan/blower assembly.

All motors used in heat pumps are *induction motors*. Current in the moving part of the induction motor is induced; the moving component has no connection to the source of current. The stationary part of an inductor motor is called the *stator*, and the moving part is the *rotor*. The stator windings are connected to the power source, while the rotor is mounted on the motor shaft, with the rotation of the rotor providing the motor driving power source.

As discussed in Chapter 1, the first law of thermodynamics states that energy can neither be created nor destroyed, but may be converted from one form into another. A motor receives electrical energy from the power source, but because of friction and efficiency losses, only a part of this input energy can be turned into mechanical output energy. The balance of the input energy is converted to heat energy, and unless this heat is dissipated, the temperature within the motor windings will rise until the insulation is destroyed. If a motor is kept free from contamination and physical damage, heat is the only enemy that can damage its windings.

The amount of heat produced in a motor depends both on the load and on motor efficiency. As the load is increased, the electrical energy input to the motor increases. The percentage of the power input converted to heat in the motor depends on motor efficiency, decreasing with an increase in efficiency and increasing as efficiency decreases.

The temperature level that a motor can tolerate depends largely on the type of motor insulation and the basic motor design, but the actual motor life is determined by the operating conditions to which it is subjected. If operated in a proper environment at loads within its design capabilities, a well-designed motor should have an indefinite life. Continuous overloading of a motor, resulting in consistently high operating temperatures, will materially shorten its life.

The great majority of motors used to power compressors in heat pumps are hermetic, meaning the motor is mounted directly on the compressor crankshaft and hermetically sealed within the compressor body. Aside from the economics inherent in this type of construction, the greatest advantage is that the motor can be cooled by a variety of means, such as air, water, or refrigerant vapor. Sealing the motor in the compressor body eliminates the troublesome problem of sealing the crankshaft so that the power may be transmitted without refrigerant leaks.

Although there are different types of electric motors, all are based on the same two electromagnetic principles: i.e., that current flow in a conductor will produce a magnetic field surrounding a conductor; and that, if a conductor is moved through a magnetic field, current is induced in it. This induced current will then create its own magnetic field. The first principle applies to the magnetic field created by the stator, and the second applies to the rotor as it rotates within the stator field.

Referring to Figure 3.37, the fundamental motor consists of: a frame, or stator, containing the stationary winding to which the voltage supply is connected; a rotor containing the secondary winding in which the current is induced by interaction with the magnetic field established by the stator winding; and endshields containing a bearing system to support the rotor and maintain its axial and radial alignment relative to the stator. The stator winding, which consists of coils of copper or aluminum magnet wire placed in a stator constructed from a stack of thin steel laminations, creates a magnetic field that is transmitted across an air gap and into the rotor. The rotor consists of a shorted aluminum winding that is cast into slots (bars) in a stack of thin steel laminations and joined at both ends of the stack with cast and rings.

Before entering into a discussion of the characteristics of the various types of electric motors, it may be helpful to discuss the basic electromagnetic phenomena that enable the induction motor to convert

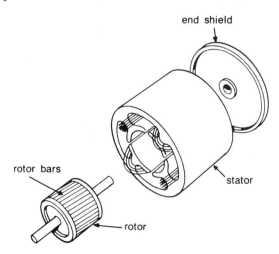

Figure 3.37   Basic components of four-pole, shaded-pole motor.

electrical energy into mechanical energy that is output through the motor shaft. Whenever current flows in a conductor, a magnetic field is built up around it, as shown in Figure 3.38. If the conductor is formed into a coil, the magnetic field created is similar to that of a permanent bar magnet, as shown in Figure 3.39. If a bar of magnetic material, such as iron or steel, is placed within the coil, the magnetic field is strengthened, because these materials transmit magnetic flux much more readily than air.

Figure 3.40 shows a view of one-half of the stator shown in Figure 3.37. Note that the placement of the coils within the stator closely resembles the relative positioning of the coil and bar of Figure 3.40, and that the resulting magnetic fields are also similar. By reversing the direction in which one coil is wound in the stator relative to the adjacent coil, the direction of current flow in that coil is reversed, as shown in

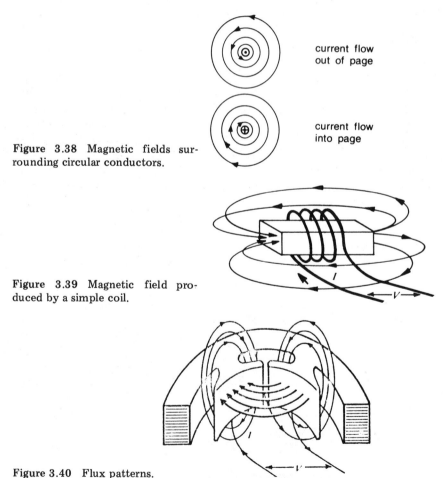

current flow
out of page

current flow
into page

Figure 3.38 Magnetic fields surrounding circular conductors.

Figure 3.39 Magnetic field produced by a simple coil.

Figure 3.40 Flux patterns.

Figure 3.40. The reversal in the direction of current flow also changes the magnetic polarity, creating adjoining north and south poles in the stator.

In the completed stator, then, a magnetic field is created having alternating north and south poles. The number of magnetic poles in the stator, together with the alternating current line frequency, determines the speed at which the motor will operate. The purpose of the rotor is to interact with this stator magnetic field and convert the potential energy stored there into useful mechanical energy; i.e., torque output at the shaft.

If a conductor is moved between the faces of a horseshoe magnet (Figure 3.41), a voltage is induced in that conductor; if this conductor forms part of a completed circuit, current will flow. The action which produces this voltage and current is the cutting of the magnetic field (the so-called *lines of magnetic flux*) by the conductor. This current flow within the conductor will, in turn, produce a magnetic field, as illustrated in Figure 3.38. The interaction of these two magnetic fields produces a mechanical force on the wire that, in the case of a direct current voltage source, would resist the movement of the conductor between the poles. The phenomenon illustrated is the basis for the production of useful torque output in AC (alternating current) induction motors.

In an induction motor, the single conductor of the previous example is replaced by the rotor winding. In small motors, this winding is normally of cast aluminum and consists of multiple conductors, or rotor bars, cast into the slots in the rotor core and joined at both ends of the stack with endrings. The resulting rotor winding circuit is shown in Figure 3.42; for obvious reasons, rotors constructed in this manner have come to be called *squirrel cage rotors*.

Consider the squirrel cage rotor centered in the energized four-pole stator shown in Figure 3.40 and rotating at its normal operating

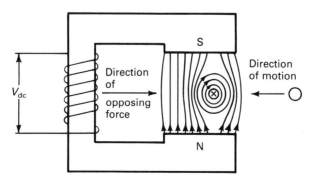

Figure 3.41   Elements of motor action.

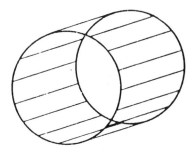

Figure 3.42 Squirrel cage rotor winding.

speed of 1550 rpm (revolutions per minute). As the rotor turns, the rotor bars cut through the flux lines of the stator magnetic field, and voltages and currents are induced in each bar. The magnitude of the voltage and current in a given bar at any instant in time will depend on the magnetic density (strength) of the stator field through which the bar is cutting at that time. Because the strength of the stator field varies around its circumference, being strongest at the center of each pole, the voltages and currents induced in the rotor bars will also vary around the rotor diameter. Figure 3.43 illustrates the directions and magnitudes of the rotor-bar currents generated at an instant in time as the rotor rotates in the four-pole stator. The magnetic field built up around each bar reacts with the stator flux, exerting a force on each bar in the same manner as previously described in the example of the horseshoe magnet and simple conductor. The sum total of all forces acting on all rotor bars is the output torque of this particular motor at 1,550 rpm.

While the motor used in this example was a shaded-pole design, the same basic principles apply to other single-phase types, such as resistance split-phase, capacitor-start, and permanent split-capacitor, and to polyphase motors as well. The following discussion will explain each of the types of electric motors in further detail, and summarize the relative strengths and weaknesses of each design.

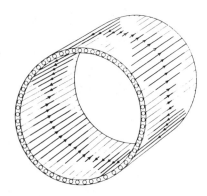

Figure 3.43 Schematic representation of squirrel cage winding containing four paths for the voltages and currents produced by the four pole stator.

Polyphase Motors

The polyphase motor is widely used in integral heat pump industrial applications, but is covered here because the starting and running performance of single-phase motors, which are used in a large majority of heat pumps, relates directly to two-phase polyphase characteristics.

Polyphase motors are distributed-wound motors; that is, the stator contains many slots, and each pole of the winding spans several slots and consists of a number of coils. One coil side is placed in each slot, and the number of turns in each coil is varied according to its position in the pole so that the turns in the pole (and therefore the magnetic flux) vary across the span of the pole. Figure 3.44(a) depicts the construction of a distributed-wound, two-pole, two-phase motor having two identical windings placed 90 degrees apart. Figures 3.44 (b) and (c) illustrate the coil arrangement and resulting magnetic flux distribution of one pole of one phase of this two-phase motor.

For induction motors to produce torque, the bars of the rotor winding must be cut by the magnetic flux lines of the stator field. In polyphase motors, this cutting action is assured at all speeds because the magnetic field of the stator itself revolves. Therefore, torque is produced at all rotor rpms, even at standstill. This revolving field is created by the supply voltages that are connected to the stator windings. In a two-phase motor, these voltages are 90 electrical degrees out of phase with each other. Because the two phase windings are identically wound and placed, the phase currents will also be 90 degrees out of phase, as illustrated in Figure 3.45. In three-phase systems, the phase voltages and windings currents are separated by 120 electrical degrees.

Figure 3.46, which depicts a two-pole, two-phase polyphase motor, illustrates how this phase shift creates a rotating magnetic field. Note in Figure 3.44 that at point *a*, phase II is "off" and only phase I is supplying power to the motor. Therefore, the poles of phase I are very strong magnetically, and flux flows across the air gap and through the rotor from the north pole to the south pole of phase I. At point *b*, phase I has weakened, but phase II is now supplying an equal amount of power to the second set of stator windings. The result is that the north magnetic pole is centered between the north poles of phase I and phase II, and the magnetic field has rotated 45 degrees. At point *c*, phase I is off and phase II transmits all of the flux. Therefore, the magnetic field has now rotated 90 degrees.

Note that in this two-pole motor, the magnetic field makes one complete revolution with every cycle of the supply voltage. On 60 Hz (Hertz) power systems, there are 3,600 cycles per minute, so the magnetic field revolves at 3,600 rpm, the so-called *synchronous speed of*

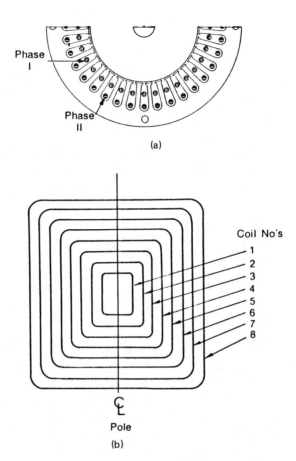

(a)

(b)

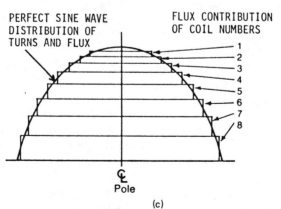

(c)

Figure 3.44 (a) Distributed winding construction; (b) Coil arrangement of one pole; (c) Resulting flux distribution.

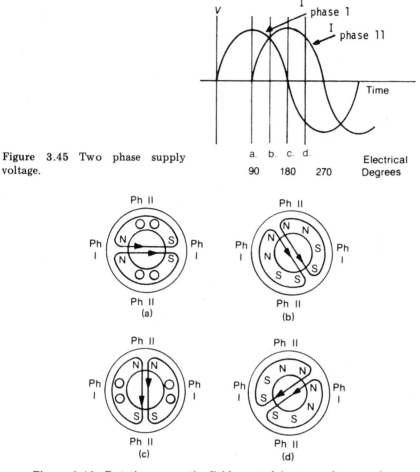

Figure 3.45 Two phase supply voltage.

Figure 3.46 Rotating magnetic field created in two-pole, two-phase motor.

the motor. In a four-pole motor, it takes two cycles of the supply voltage to rotate the magnetic field once. Therefore, the synchronous speed is 1,800 rpm. In general, synchronous rpm is equal to the frequency multiplied by 120, which is then divided by the number of poles.

The speed/torque relationship of a typical polyphase design is illustrated in Figure 3.47 and general polyphase characteristics are summarized in Table 3.1. Use of polyphase designs is limited for the most part to larger, industrial-type motors, because of the relatively high cost of multiphase power systems and the motors themselves. To serve the domestic market, several types of less expensive designs have evolved that operate from a single AC voltage supply: single-phase motors.

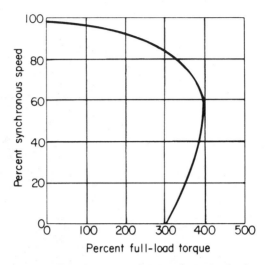

Figure 3.47   General performance characteristics, polyphase design.

TABLE 3.1   Polyphase Characteristics

Characteristic	Rating	Comment
Efficiency	High	60–80%
Power Factor	Moderate to High	70–80%
Starting Torque	High	100–300%
Noise and Vibration	Low	No 120 Hz torque pulsations
Cost	High	

Single-phase Motors

In polyphase motors, the correct mechanical placement of the multiple windings in the stator, together with the phase relationships between the supply voltages, produce a uniformly rotating stator magnetic field. With a single source of AC voltage connected to a single winding, however, this is not the case. A stationary flux field is created that pulsates in strength as the AC voltage varies, but does not rotate. Consequently, if a stationary rotor is placed in this single phase stator field, it will not rotate. If the rotor is spun by hand, artificially creating relative motion between the rotor winding and the stator field, it will pick up speed and run. The fact is that a single-phase, single-winding motor will run (in either direction) if started by hand, but will not develop any starting torque (see Figure 3.48).

What is needed is a second winding with currents out of phase with the original, or main, winding to produce a net rotating magnetic field, as was the case with the two-phase polyphase motor. The various

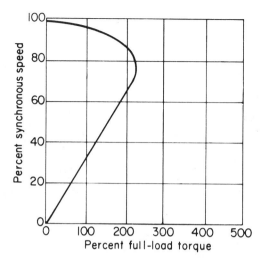

Figure 3.48   Speed-torque characteristics of single-phase, single-winding motor.

single-phase motor designs differ in the type of secondary, or start, winding employed. These start windings, which together with components such as capacitors, relays, and centrifugal switches make up the starting circuit, have varying effects on motor starting and running performance.

Two different, but equivalent mathematical models for single-phase motors have been developed: the double revolving theory and the cross field theory. These theories will not be explored here, but both reveal two important facts about single-phase motors:

1. The performance (speed/torque characteristics, efficiency, and so on) of single-phase motors approaches that of the two-phase polyphase motor as an upper limit.

2. The torque produced at a given rpm by a single-phase motor is not constant, but pulsates at twice the line frequency (120 Hz on a 60 Hz system) around a median value. These torque pulsations are inherent to all single-phase motors and can cause noise and vibration problems if not properly isolated.

Resistance Split-phase (KH) Motors

Figure 3.49 illustrates schematically the winding arrangement of a typical distributed winding resistance split-phase motor. If only the main winding is energized and the main winding current is recorded, the relationship between the supply voltage (V_L) and the main winding current (I_M) would be as shown in Figure 3.50(a). The main winding cur-

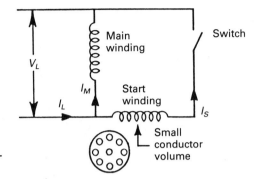

Figure 3.49 Winding schematic, resistance split-phase motor.

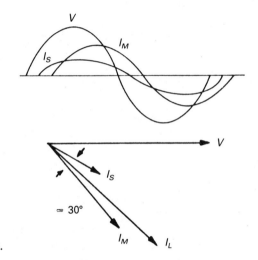

Figure 3.50 Phase relationships.

rent lags behind the line voltage, because the coils embedded in the steel stator naturally build up a strong magnetic field which slows the build up of current in the winding. The relationship between the line voltage and line current is shown in Figure 3.49(b).

The start winding is not wound identical to the main, as was the case with the two-phase, polyphase motor, but its coils contain fewer turns of much smaller diameter wire than the main winding coils. This is required to reduce the amount the start-winding current lags the voltage if it is connected to the line, as shown in Figure 3.49(a) and (b).

When both windings are connected in parallel across the line, the main and start winding currents will then be out of time phase, not by 90 degrees (as in the two-phase system), but by about 30 degrees. This forms a sort of imitation two-phase system, which produces a weak rotating flux field that is nevertheless sufficient to provide a moderate amount of torque at standstill and start the motor.

The total current that the motor draws when starting is the vector sum of the main and start-winding currents. Because of the small angle between these two vectors, the line current drawn during starting (inrush current) of split-phase motors is quite high. Also, the small diameter wire in the start winding carries a high current density, so it heats up very quickly. A centrifugal switch and mechanism or relay must be provided to remove the start winding from the circuit once the motor has reached an adequate speed to allow running on the main winding only. Figure 3.51 illustrates the speed/torque relationship of a typical split-phase motor on both the running and starting connection.

Running on the main winding only, the split-phase motor is subject to 120 Hz torque pulsations and does not achieve as high an efficiency level as the two-phase, polyphase motor. Split-phase motor characteristics are summarized in Table 3.2.

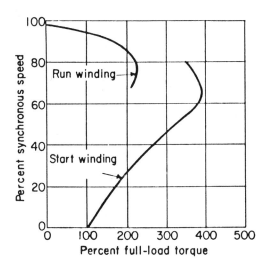

Figure 3.51   General performance characteristics, split-phase design.

TABLE 3.2   Split-phase Motor Characteristics

Characteristic	Rating	Comment
Efficiency	Moderate	50–65%
Power Factor	Moderate	60–70%
Starting Torque	Moderate	High inrush current starting
Noise & Vibration	Moderate	120 Hz torque pulsations
Cost	Moderate	Self-cooled, contains switch & mechanism

Capacitor-start (KC) Motors

The capacitor-start motor utilizes the same winding arrangement as the split-phase motor (Figure 3.52), but adds a short-time rated capacitor in series with the start winding. (A *capacitor* is an electrical device that is able to store electrical energy. Capacitors will be discussed in detail in Section 3.5.)

The effect of the addition of this capacitor is illustrated in Figure 3.53. The main winding current (I_M) remains the same as in the split-phase case, but the start-winding current is very different. Because of the addition of the capacitor, it now leads the line voltage rather than lagging, as does the main winding. The start winding itself is also different, containing slightly more turns in its coils than the main winding, and utilizing wire diameters only slightly smaller than those of the main.

The net result is a time phase shift that is much closer to 90 electrical degrees than with the split-phase motor. A stronger rotating field is therefore created, and starting torque is higher than with the split-phase design. Note also that the vector sum of the main and start-winding cur-

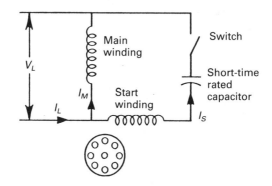

Figure 3.52   Winding schematic, capacitor start motor.

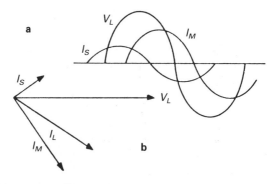

Figure 3.53   Phase relationships, capacitor start motor.

rents is substantially lower, resulting in greater locked rotor torques at reduced inrush currents.

The starting and running speed/torque characteristics of a typical capacitor start motor are illustrated in Figure 3.54. Again, a centrifugal switch and mechanism or relay must be used to protect both the start winding and capacitor from damage due to overheating. When running, the capacitor start motor performs identically to the split-phase motor. Table 3.3 summarizes capacitor start motor characteristics.

Permanent Split-capacitor (KCP) Motors

The windings of the permanent split-capacitor motor (Figure 3.55) are arranged like those of the split-phase and capacitor-start designs. A capacitor capable of running continuously replaces the intermittent-duty capacitor of the capacitor-start motor and the centrifugal switch is removed. Once again, the main-winding current remains similar to the previous designs, lagging the line voltage (Figure 3.56). The start-winding

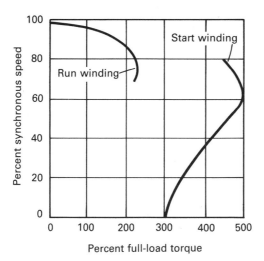

Figure 3.54 General performance characteristics, capacitor start design.

TABLE 3.3 Capacitor-start Motor Characteristics

Characteristic	Rating	Comment
Efficiency	Moderate	50–65%
Power Factor	Moderate	60–70%
Starting Torque	Moderate to High	Capacitor controls inrush current
Noise & Vibration	Moderate	120 Hz torque pulsations
Cost	Moderate	Self-cooled, contains switch & mechanism or relay

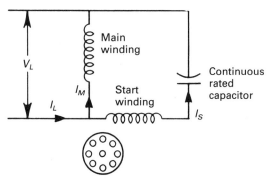

Figure 3.55 Winding schematic, permanent split-capacitor motor.

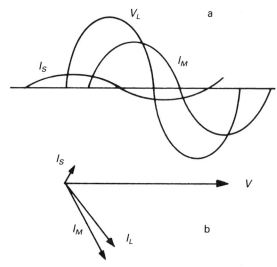

Figure 3.56 Phase relationships, permanent split-capacitor motor.

current and the start winding itself, however, are somewhat different than in the capacitor-start design.

Continuously rated capacitors are normally provided in small microfarad, high-voltage ratings. Therefore, the start winding is altered to boost the capacitor voltage to the correct level by adding considerably more turns to its coils than are in the main winding coils. Start-winding wire size remains somewhat smaller than that of the main windings. The smaller microfarad rating of the capacitor produces more of a leading phase shift and less total start-winding current, so starting torques will be considerably lower than with the capacitor-start design.

However, the real strength of the permanent split-capacitor design is derived from the fact that the start winding and capacitor remain in the circuit at all times and produce an approximation of two-phase, polyphase operation at the rated load point. This results in better efficiency, better power factor, and lower 120 Hz torque pulsations than in equivalent capacitor-start and split-phase designs. With modern computer optimization techniques, a winding design and capacitor selection can be made that will result in efficiency equal to a two-phase, polyphase motor at one (and only one) given load point. With a perfect 1.0 power factor, the permanent split-phase motor is better than a polyphase motor. It is for this reason that many heat pump manufacturers now incorporate the permanent split-capacitor type of motor in their heat pumps.

Figure 3.57 illustrates typical speed/torque curves for permanent split-capacitor motors. Note that varying the rotor resistance (size of the rotor bars and endrings) changes the shape of the speed/torque curve so that various operating rpms can be utilized. In addition, by adding extra main windings in series with the original main windings, motors can be designed to operate at different speeds depending on the number of extra main windings energized. Note also that for a given full load torque, less breakdown torque, and therefore a smaller motor, is required with a permanent split-capacitor design than with any of the previously discussed designs. Table 3.4 summarizes the characteristics of the permanent split-capacitor design.

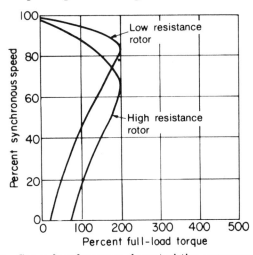

Figure 3.57   General performance characteristics, permanent split-phase design.

TABLE 3.4   Permanent Split-capacitor Motor Characteristics

Characteristic	Rating	Comment
Efficiency	Moderate to High	50–70%
Power Factor	High	80–100%
Starting Torque	Low to Moderate	Varies with rotor resistance
Noise & Vibration	Moderate to Low	120 Hz torque pulsations reduced
Cost	Moderate to Low	No switch & mechanism or relay smallest motor for given output

Shaded-pole (KSP) Motors

The shaded-pole motor greatly differs from the other single-phase designs. All of these designs were distributed-wound motors having a main and a start winding, differing only in details of the starting method and corresponding starting circuitry. The shaded-pole motor is entirely different, both in construction and operation.

The shaded-pole motor is the most simply constructed and least expensive of the single-phase designs. It consists of a run winding only, plus shading coils that take the place of a conventional start winding. Figure 3.58 illustrates the construction of a typical four-pole, shaded-pole motor. The stator is of salient pole-pole construction, having one large coil per pole wound directly in a single large slot. The shading coils are short-circuited copper straps that are wrapped around one pole tip of each pole.

The shaded-pole motor produces a very crude approximation of a rotating stator field through magnetic coupling, which occurs between the shading coils and the stator winding. The placement and resistance of the shading coil is chosen so that, as the stator magnetic field increases from zero at the beginning of the AC cycle to some positive value, current is induced in the shading coil. As previously noted, this current will create its own magnetic field which opposes the original field. The net effect is that the shaded portion of the pole is weakened and the magnetic center of the entire pole is located at point a. As the flux magnitude becomes nearly constant across the entire pole tip at the top of the positive half-cycle, the effect of the shading coil is negligible and the magnetic center of the pole shifts to point b. Although this shift is slight, it is sufficient to generate torque and start the motor.

Figure 3.59 illustrates the effect of the salient pole-winding configuration on motor speed/torque output. As has been pointed out previously, a sinusoidal distribution of winding turns produces the best efficiency and lowest noise and vibration levels. The single coil winding

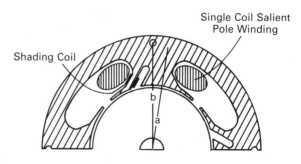

Figure 3.58   Four-pole, shaded-pole motor.

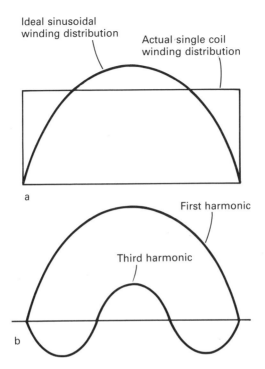

Figure 3.59 Harmonic content resulting from single coil winding distribution.

is the crudest possible approximation of such a distribution. Therefore, shaded-pole motor efficiency suffers greatly due to the presence of winding harmonic content, particularly the third harmonic, which produces a dip in the speed/torque curve at approximately one-third of synchronous speed (see Figure 3.60). In addition, there are always losses present in the shading coils.

These factors combine to make the shaded-pole motor the least efficient and noisiest of the single-phase designs. It is used primarily in air-moving applications where its low starting torque can be tolerated. Extra main windings can be added to provide additional speeds in a manner similar to that used on permanent split-capacitor motors. Table 3.5 summarizes shaded-pole motor characteristics.

3.5 CAPACITORS

A *capacitor* is an electrical device that stores electrical energy. Capacitors are used in electric motors primarily to displace the phase of the current passing through the start winding. While a detailed study of

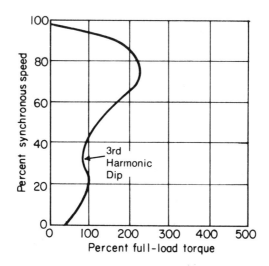

Figure 3.60 General performance characteristics, four-pole, shaded-pole design.

TABLE 3.5   Shaded-pole Motor Characteristics

Characteristic	Rating	Comment
Efficiency	Low	20–40%
Power Factor	Low	50–60%
Starting Torque	Low	Plus 3rd harmonic dip
Noise & Vibration	High	120 Hz plus winding harmonics
Cost	Low	

electrical theory is beyond the scope of this book, capacitors in a motor circuit provide starting torque, and improve running characteristics, efficiency, and the power factor.

The amount of electrical energy a capacitor will hold depends on the voltage applied. If the voltage is increased, the amount of electrical energy stored in the capacitor is increased. The capacity of a capacitor is expressed in microfarads (MFD), and is dependent on the size and construction of the capacitor.

The voltage rating of a capacitor indicates the nominal voltage at which it is designed to operate. Use of a capacitor at voltages below its rating will do no harm. Run capacitors must not be subjected to voltages exceeding 110% of the nominal rating, and start capacitors must not be subjected to voltages exceeding 130% of the nominal rating. The voltage to which a capacitor is subjected is not line voltage, but is a much higher potential (often called electromotive force or back electromotive force) that is generated in the start winding.

Capacitors, either start or run, can be connected either in series or parallel to provide the desired characteristics, if the voltage MFD are properly selected. When two capacitors having the same MFD rating are connected in series, the resulting total capacitance will be one-half the rated capacitance of a single capacitor. The formula for determining capacitance (MFD) when capacitors are connected in series is as follows:

$$\frac{1}{\text{MFD total}} = \frac{1}{\text{MFD}_1} + \frac{1}{\text{MFD}_2}$$

For example, if a 20 MFD and a 30 MFD capacitor are connected in series, the resultant capacitance will be:

$$\frac{1}{\text{MFD}_t} = \frac{1}{\text{MFD}_1} + \frac{1}{\text{MFD}_2}$$

$$\frac{1}{\text{MFD}_t} = \frac{1}{20} + \frac{1}{30}$$

$$\frac{1}{\text{MFD}_t} = \frac{5}{60} = \frac{1}{12}$$

$$\text{MFD}_t = 12 \text{ MFD}$$

The voltage rating of similar capacitors connected in series is equal to the sum of the voltage of the two capacitors. However, since the voltage across individual capacitors in series will vary with the rating of the capacitor, for emergency field replacements it is recommended that only capacitors of like voltage and capacitance be connected in series, to avoid the possibility of damage due to voltage beyond the capacitors' limits.

When capacitors are connected in parallel, their MFD rating is equal to the sum of the individual ratings. The voltage rating is equal to the smallest rating of the individual capacitors. It is possible to use any combination of single, series, or parallel starting capacitors with single or parallel running capacitors, although running capacitors are seldom used in series.

Start capacitors are designed for intermittent service only and have a high MFD rating. Their construction is of the electrolytic type, in order to obtain higher capacity. Start capacitors are normally supplied with bleed resistors securely attached and soldered to their terminals, as shown in Figure 3.61. The use of capacitors without bleeder resistors can result in sticking relay contacts and/or erratic relay operation. This is due to the starting capacitor discharging through the relay contacts as they close, following a very short running cycle. The resistor will permit the capacitor charge to bleed down at a much faster rate, preventing arcing and overheating of the relay contacts.

Run capacitors are continuously in service in the operating circuit

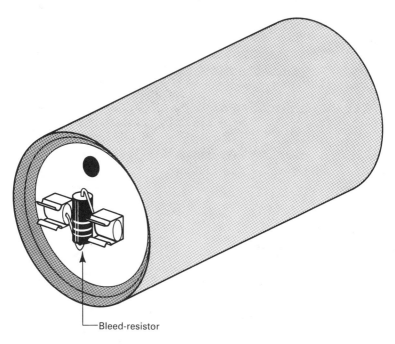

Figure 3.61 Starting capacitor with bleed resistor (Courtesy Copeland Corporation).

and are normally of the oil-filled type. The run capacitor capacitance rating is much lower than a start capacitor. Because of the voltage generated in the motor start winding, the run capacitor has a voltage across its terminals greater than the line voltage.

The starting winding of a motor can be damaged by a shorted and grounded running capacitor. This damage usually can be avoided by proper connection of the run-capacitor terminals. The terminal connected to the outer foil (nearest the can) is the one most likely to short to the can and be grounded in the event of a capacitor breakdown. It is identified and marked by most manufacturers of run capacitors (See Figure 3.62).

From the supply line of a typical 115 or 230 volt circuit, a 115 volt potential exists from the R (run) terminal to ground through a possible short in the capacitor (as shown in the wiring diagram of Figure 3.63). However, from the S or start terminal, a much higher potential, possibly as high as 400 volts, exists because of the electromotive force (EMF) generated in the start winding. Therefore, the possibility of capacitor failure is much greater when the identified terminal is connected to the S terminal.

The identified terminal should always be connected to the supply line, or R terminal, never to the S terminal. If connected in this manner,

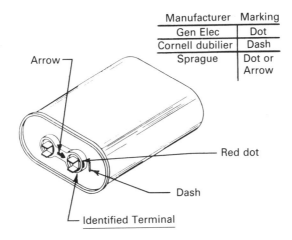

Manufacturer	Marking
Gen Elec	Dot
Cornell dubilier	Dash
Sprague	Dot or Arrow

Arrow

Red dot

Dash

Identified Terminal

Figure 3.62   Run capacitor (Courtesy Copeland Corporation).

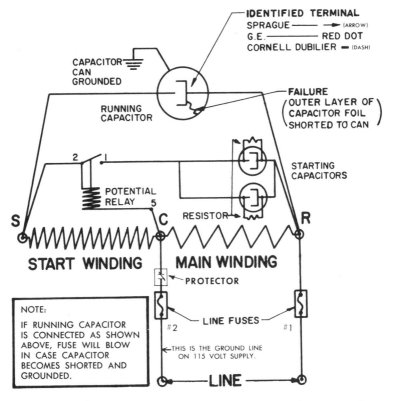

IDENTIFIED TERMINAL
SPRAGUE ———— ➡ (ARROW)
G.E. ———————— RED DOT
CORNELL DUBILIER ➡ (DASH)

CAPACITOR
CAN
GROUNDED

RUNNING
CAPACITOR

FAILURE
(OUTER LAYER OF
CAPACITOR FOIL
SHORTED TO CAN)

2   1

STARTING
CAPACITORS

POTENTIAL
RELAY   5

RESISTOR

S   C   R

START WINDING   MAIN WINDING

PROTECTOR

NOTE:

IF RUNNING CAPACITOR
IS CONNECTED AS SHOWN
ABOVE, FUSE WILL BLOW
IN CASE CAPACITOR
BECOMES SHORTED AND
GROUNDED.

LINE FUSES

#2   #1

←THIS IS THE GROUND LINE
ON 115 VOLT SUPPLY.

LINE

Figure 3.63   Wiring diagram of motor with start and run capacitors
(Courtesy Copeland Corporation).

a shorted and grounded run capacitor will result in a direct short to ground from the R terminal and will blow line fuse No. 1. The motor protector will protect the main winding from excessive temperature. If, however, the shorted and grounded terminal is connected to the start winding terminal S, current will flow from the supply line through the main winding and through the start winding to ground. Even though the protector may trip, current will continue to flow through the start winding to ground, resulting in a continuing temperature rise and failure of the starting winding.

Four
Tools, Instruments, and Standard Service Procedures

The heat pump technician will need a satisfactory complement of tools and measuring and testing instruments to perform the many and varied installation and service procedures associated with heat pumps. Although each technician's complement may vary somewhat depending on individual preference, this chapter will present those most commonly used and, where applicable, describe standard service procedures associated with the various types of tools and test equipment.

4.1 TOOLS

The following tools should be considered standard in the heat pump technician's tool box:

- **Tube cutter:** used for cutting copper tubing
- **Hack saw:** may also be used for cutting copper tubing, although a tube cutter is recommended
- **Flaring tool:** used to join copper tubing
- **Soldering equipment**
- **Wrenches of various sizes and types**

- Pliers of various sizes and types
- Screwdrivers of various sizes and types
- Hammers of various sizes and types
- Files of various sizes and types
- Tape measure and hand rule
- Micrometer and calipers
- Drills of various sizes and types

4.2 REFRIGERANT MEASURING AND TESTING EQUIPMENT

In addition to the above, the heat pump technician will also need the appropriate measuring and testing equipment that is associated with the use of refrigerants. This equipment is used to determine refrigerant temperature and pressure conditions. Each type will be discussed individually. The following sections (pp. 96-114) have been excerpted from *Refrigeration and Air-Conditioning* (see footnote, p. 114).

Temperature Measurements

When analyzing a heat pump system, accurate temperature readings are important. The most common temperature measuring device is the pocket glass thermometer, illustrated in Figure 4.1.

The thermometer head has a ring for attaching a string to suspend it, if needed. The ranges of glass thermometers vary, but $-30°$ to $+120°F$ is a common scale, with calibrations in $2°$ marks. Some have a mercury fill, but others use a red fill that is easier to read.

To check the calibration of a pocket glass thermometer, insert it into a glass of ice water for several minutes. It should read $32°F$, plus or minus $1°$. Should the fill separate, place the thermometer in a freezer, and the resulting contraction will probably rejoin the separated column of fluid. Another way to connect a separated fill is to carefully heat the stem, not the bulb. In most cases, the liquid will coalesce as it expands.

Another form of pocket thermometer is the dial type, shown in Figure 4.2. It also has a carrying case with a pocket clip. The dial thermometer is more convenient and practical for measuring air temperatures in a duct. The stem is inserted into the cut, but the dial remains visible. Again, several ranges are available, depending on the accuracy needed and the nature of application. A common range is $-40°F$ to $+160°F$.

A different type of dial thermometer is the superheat thermometer, as

Figure 4.1   Pocket glass thermometer.

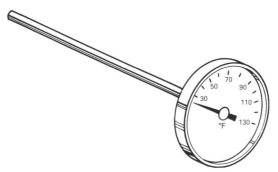

Figure 4.2 Dial-type pocket thermometer.

illustrated in Figure 4.3. The highly accurate expansion-bulb thermometer is used to measure suction line temperature(s) in order to calculate, check, and adjust superheat. A common range is $-40°$ F to $+65°$ F. The sensing bulb is strapped or clamped to the refrigerant line and then covered with insulating material (such as a foam rubber sheet) to prevent air circulation over the bulb while taking a reading. Of course, the superheat bulb thermometer can be used to measure air or water tempratures as well.

These thermometers have been the basic temperature-measuring tools

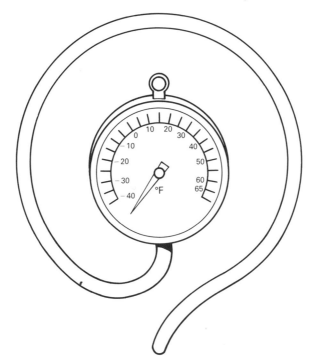

Figure 4.3 Superheat thermometer.

for many years. However they do have certain limitations; for example, the operator is required to actually be in the immediate area during the time of reading.

For remote use, some thermometers have probe lengths of up to 30 inches. Different probes are available to measure surface or air temperatures. In heat pump use, the surface probe determines superheat settings of expansion valves, motor temperature, evaporating and condensing temperatures, and water temperatures. This type of thermometer is a useful instrument for many applications, but it too is limited, since only one reading in one place at a time can be taken.

As a result of the rapid development of low-cost electronic devices, the availability and use of electronic thermometers is now very common. The electronic thermometer consists of a tester with provision to attach one or several (three to six) sensing leads. The sensing lead tip is actually a thermistor element, which, when subject to heat or cold, will vary the electrical current in the test circuit because its resistance changes with temperature change. The tester converts changes in electrical current to temperature readings. The sensing leads vary in length depending on the make of the unit, but extensions can be used to allow for remote testing.

Once the operator places the sensing probe(s) in the spots chosen to be tested, he or she may switch from position to position and record temperatures without actually entering each test area. The sensing probe can also be used to check superheat.

Pressure Measurements

Temperature measurements are usually taken outside of the operating system. But it is also necessary for the service technician to know what is going on inside of the system; this is learned from pressure measurements.

A *manometer* is one type of device utilized for the measurement of pressure. This type of pressure gauge utilizes a liquid—usually mercury, water, or gauge oil—as an indicator of the amount of pressure involved. The water manometer or water gauge is customarily used when measuring air pressures because of the lightness of the fluid being measured.

A simple open-arm manometer is shown in Figure 4.4(a) and (b). The U-shaped glass tube is partially filled with water, as shown in Figure 4.4(a), and is open at both ends. The water is at the same level in both arms of the manometer, because both arms are open to the atmosphere and there is no external pressure being exerted on them.

Figure 4.4(b) shows the manometer in use with one arm connected to a source of positive air pressure that is being measured. The water is at different levels in the arms, and the difference denotes the amount of pressure being applied.

A space that is void, or lacking any pressure, is described as having a *perfect vacuum*. If the space's pressure is less than atmospheric pressure, it is defined as being in a *partial vacuum*. It is customary to express this partial vacuum in inches of mercury, not as negative pressure. In some instances, it is

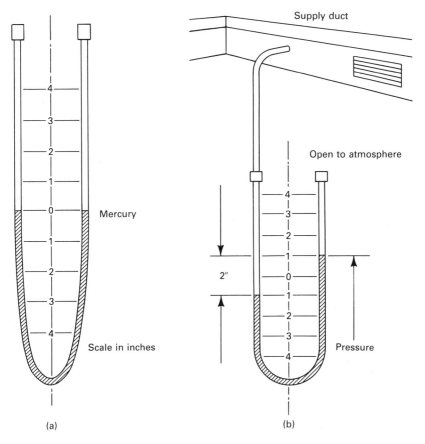

Figure 4.4   Water fill manometer. From Air-Conditioning and Refrigeration Institute, *Refrigeration and Air-Conditioning*, © 1979, p. 29 Reprinted by permission of Prentice-Hall, Inc.

also referred to as a given amount of absolute pressure, expressed in psia (pounds per square inch absolute).

If a partial vacuum has been drawn on the left arm of the manometer by means of a vacuum pump, as shown in Figure 4.5, the mercury in the right arm will be lower, and the difference in levels will designate the partial vacuum in inches of mercury.

The most common pressure gauges utilized by service technicians in the field to determine what is happening in the refrigeration system are of the Bourdon-tube type. As shown in Figure 4.6 (an internal view of this gauge) the essential element of this type of gauge is the Bourdon tube. This oval metal tube is curved along its length and forms an almost complete circle. One end of the tube is closed, and the other end is connected to the equipment or component being tested.

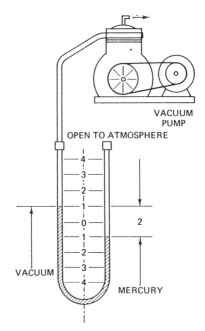

Figure 4.5 Mercury fill manometer. From Air-Conditioning and Refrigeration Institute, *Refrigeration and Air-Conditioning*, © 1979, p. 29 Reprinted by permission of Prentice-Hall, Inc.

As shown in Figure 4.7(a) and (b), the gauges are preset at 0 pounds, which represents the atmospheric pressure of 14.7 psi. Therefore, any additional pressure applied when the gauge is connected to a piece of equipment will tend to straighten out the Bourdon tube, thereby moving the needle or pointer and its mechanical linkage, thus indicating the amount of pressure being applied. Figure 4.7(a) is an example of a pressure gauge that indicates the amount of pressure above the atmospheric level. Figure 4.7(b) is a compound pressure gauge, which has a dual function: it registers a pressure above atmospheric pressure, and one that may be below atmospheric pressure.

Figure 4.7(b) shows a high-pressure gauge, which measures high-side or condensing pressures. It is normally graduated from 0 to 500 psi in 5-pound graduations. The compound gauge (Figure 4.7(a)) is used on the low side (suction pressures), and is normally graduated from a 30-inch vacuum to 120 psi; thus, it can measure pressure above and below atmospheric pressure. This gauge is calibrated in 1-pound graduations. Other pressure ranges are available for both gauges, but these two are the most common.

A service device that includes both the high and compound gauges is called a *gauge manifold*. It enables the technician to check system operating pressures, add or remove refrigerant, add oil, purge noncondensibles, bypass the compressor, analyze system conditions, and perform many other operations, without replacing gauges or trying to operate service connections in inaccessible places.

The testing manifold, as illustrated in Figure 4.8, consists of a service manifold containing service valves. On the left, the compound gauge (suction) is mounted and, on the right, the high-pressure gauge (discharge). On the

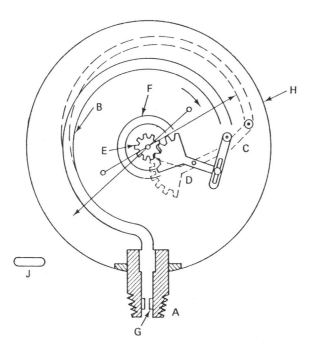

Figure 4.6 Internal construction of a pressure gauge. (a) Adapter fitting, usually an $1/8''$ pipe thread; (b) Bourdon tube; (c) Link; (d) Gear sector; (e) Pinter shaft gear; (f) Calibrating spring; (g) Restricter; (h) Case; (j) crosssection of the Bourdon tube. The dotted line indicates how the pressure in the Bourdon tube causes it to straighten and operate the gauge. From Air-Conditioning and Refrigeration Institute, *Refrigeration and Air-Conditioning*, © 1979, p. 30. Reprinted by permission of Prentice-Hall, Inc.

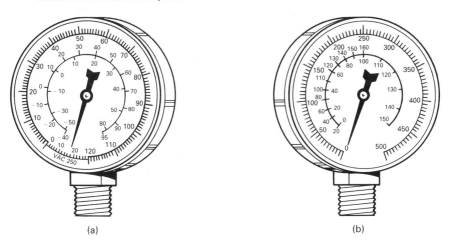

(a)   (b)

Figure 4.7 Compound gauge (left) and high-pressure gauge (right) (Courtesy Copeland Corporation).

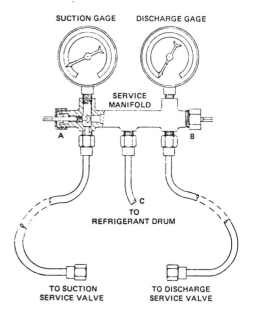

Figure 4.8 Testing manifold. From Air-Conditioning and Refrigeration Institute, *Refrigeration and Air-Conditioning*, © 1979, p. 105 Reprinted by permission of Prentice-Hall, Inc.

bottom of the manifold are hoses that lead to the equipment suction service valve (left), refrigerant drum (middle), and the equipment discharge, or liquid-line valve (right).

Many equipment manufacturers give a color code of blue to the low-side gauge casing and hose b and red to the high-side gauge and hose r. The center or refrigerant hose is colored white. This system is very helpful to avoid crossing hoses and damaging gauges. A hook is provided to hang the assembly, freeing the operator from holding it. By opening and closing the refrigerant valves on gauge manifold A and B (Figure 4.9), we can obtain different refrigerant flow patterns. The valving is so arranged that when the valves are closed (front-seated), the center port on the manifold is closed to the gauges (Figure 4.9(A)). When the valves are in the closed position, gauge ports 1 and 2 are still open to the gauges, permitting the gauges to register system pressures.

With the low-side valve (1) open and the high-side (2) closed (Figure 4.9(b)), the refrigerant is allowed to pass through the low side of the manifold and the center-port connection. This arrangement might be used when refrigerant or oil is added to the system.

Figure 4.9(c) illustrates the procedure for bypassing refrigerant from the high side to the low side. Both valves are open and the center port is capped. Refrigerant will always flow from the high-pressure area to a lower-pressure area.

Figure 4.9(d) shows the valving arrangement for purging or removing refrigerant. The low-side valve is closed. The center port is open to the atmosphere or connected to an empty refrigerant drum. The high-side valve is opened, permitting a flow of high pressure out of the center port.

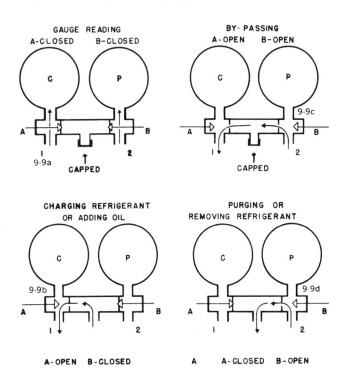

Figure 4.9   Manifold valve operation. C:compound gauge; P:pressure gauge; 1:gauge line to suction service; 2:gauge line to discharge service. From Air-Conditioning and Refrigeration Institute, *Refrigeration and Air-Conditioning*, © 1979, p. 106. Reprinted by permission of Prentice-Hall, Inc.

Note: Purging large quantities of flurocarbon refrigerant to the open atmosphere is not recommended unless absolutely necessary.

The method of connecting the gauge manifold to a system depends on the state of the system; that is, whether the system is operating or just being installed. For example, let's assume the system is operating and equipped with back seating inline service valves (Figure 4.10). The first step is to purge the gauge manifold of contaminants before connecting it to the system.

1. Remove the valve stem caps from the equipment service valves and check to be sure that both service valves are back-seated.
2. Remove the gauge port caps from both service valves.
3. Connect center hose from gauge manifold to a refrigerant cylinder, using the same type of refrigerant that is in the system, and open both valves on the gauge manifold.
4. Open the valve on the refrigerant cylinder for about 2 seconds, and then close it. This will purge any contaminants from the gauge manifold and hoses.

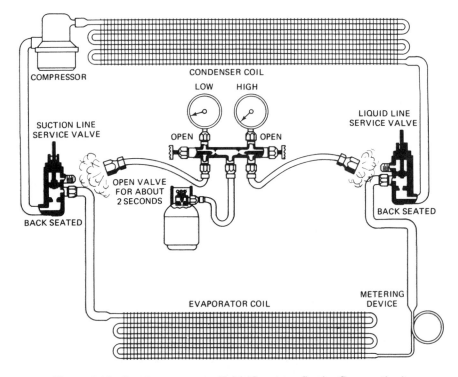

Figure 4.10   Purging gauge manifold (Courtesy Carrier Corporation).

5. Connect the gauge manifold hoses to the gauge ports: the low-pressure compound gauge to the suction service valve and the high-pressure gauge to the liquid line service valve, as illustrated in Figure 4.11.

6. Front-seat or close both valves on the gauge manifold. Crack (turn clockwise) both service valves one turn off the back seat. The system is now allowed to register on each gauge. With the gauge manifold and hoses purged and connected to the system, we are free to perform whatever service function is necessary.

To remove the gauge manifold from the system, follow this procedure:

1. Back-seat (turn counterclockwise) both the liquid and suction service valves on the in-line type.

2. Remove hoses from gauge ports and seal ends of hoses with 1/4 in. flare plugs to prevent hoses from being contaminated. (Some manifold assemblies have built-in hose seal fittings.)

3. Replace all gauge port and valve stem caps. Make sure all caps contain the gaskets provided with them and are tight.

The manifold and gauges are necessary tools to perform many system

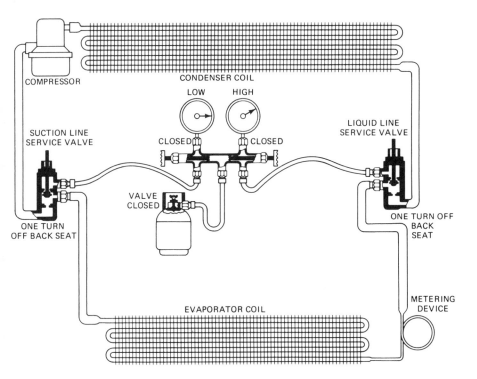

Figure 4.11   Connecting manifold (Courtesy Carrier Corporation).

operations. Once the system has been completed and cleaned of most of the air by purging, it must be tested for leaks. Or, whenever a component has been repaired or replaced, it is imperative that the entire system be checked for leaks.

4.3 LEAK TESTING

In order to check a system for leaks, it is necessary for the system or a portion of the system to be pressurized. This will naturally be true of a new system prior to evacuation and charging, or of an old system that has lost its charge. If the system has been in operation and has lost most or all of its charge, it is desirable to pressurize the entire system to find the leak or leaks. When the entire unit is to be pressurized, it's usually desirable to pressurize the system through both the suction and discharge service valves. In this manner, the pressure is supplied to the system on both sides of the capillary tube.

    Systems are commonly pressurized for the purpose of leak testing with either the refrigerant that is being used in the system or dry nitrogen, or a combination of both. However, it is recommended that the refrigerant specified by the manufacturer for the system be used for leak testing. Test pressure should be adjusted to 100 psi or higher, with the maximum being 175 psi. Refrigerant leaks can be detected with either a propane or electronic leak detector. Nitrogen cannot be detected unless it is used with a portion of the refrigerant.

To perform leak testing:

1. Attach the refrigerant cylinder to the center port of the gauge manifold. If nitrogen is used, be sure the bottle is equipped with a pressure regulator.
2. Open both valves on the gauge manifold and allow the gas to flow into the system. When refrigerant is used, don't try to pressurize beyond the saturation pressure of the refrigerant, corresponding to the ambient temperature.
3. Proceed to check for leaks using an electronic leak detector, which is the preferred and most accurate tool for leak detection. Pass the probe along the line going around the joints and connections. Since refrigerant is heavier than air, make sure that the probe tip passes on the underside of the line. The presence of a leak will be indicated by either a light, an audible sound, or both, depending on the equipment being used to test for leaks.
4. After all piping has been checked, and leaks have been located and marked, remove the refrigerant from the system or pump down the system.
5. Braze all leaks.
6. Retest the system using the above procedure.

4.4 PURGING

Whenever a system is exposed to atmospheric conditions for a short period of time during component replacement (less than 5 minutes, for example), it is necessary to purge the system to remove any contaminants that may have entered the system. Similarly, during installation, if the refrigerant lines are left open for more than 5 minutes, the system should be purged.

To purge a system that has just been installed, proceed as follows. Install the gauge manifold as illustrated in Figure 4.12, with the low-side valve closed and not connected at the suction-service valve. Connect the center hose to the refrigerant drum. Connect the high-side hose to the liquid-line valve. Front-seat both service valves and open high-side manifold valve. Open valve (wide) on the refrigerant cylinder and allow a high-velocity charge of refrigerant vapor (1/2 to 1 pound or more depending on size of equipment) to enter the system. The refrigerant will push any contaminants through the system to the suction-service valve, where they will be purged through the gauge port.

Whenever a defective component is to be removed, the system should be pumped down, and the part of the system that is to be replaced should be isolated by means of the service valves. Then, when the new component is installed, the lines should be purged from both sides.

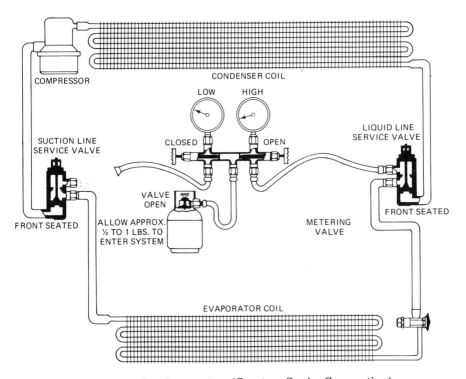

Figure 4.12   Purging a system (Courtesy Carrier Corporation).

4.5 EVACUATION

Proper evacuation of a unit will remove noncondensibles (mainly air, water, and inert gases) from the system, and assure a tight, dry system before charging. There are generally two methods used to evacuate a system: the *deep-vacuum* method, and the *triple-evacuation* method. Each has its advantages and disadvantages. The choice depends on several factors: type of vacuum pump available, the time that can be spent on the job, and whether there is liquid water in the system.

In systems that operate at very low suction pressures, the deep-vacuum method is recommended. In higher temperature systems and in air-conditioning work, triple evacuation is practiced. Both methods will be discussed.

The tools needed to evacuate a system properly depend on the method used. A good vacuum pump and vacuum indicator are needed for the deep-vacuum method, and a good vacuum pump and compound gauge are needed for the triple-evacuation method.

Vacuum Pump

A vacuum pump, as illustrated in Figure 4-13, functions like an air compressor operating in reverse. Most vacuum pumps are driven by a direct or belt-driven electric motor, but gasoline engine-driven pumps are also available. The pump

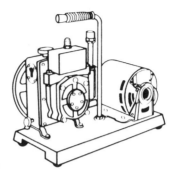

Figure 4.13 Vacuum pump.

may be single or two stage depending on the design. Most pumps for normal field service are portable; they have carrying handles or are mounted on wheel dollies. Vacuum pump sizes are rated according to their free-air displacement in cubic feet per minute or liters per minute metric. Specifications may also include a statement as to the degree of vacuum the pump can achieve, expressed in terms of microns.

What is a micron? When the vacuum pressure approaches 29.5 in. to 30 in. on the compound gauge, the gauge is working within the last half inch of pressure, and the readout beyond 29.5 in. is not reliable for the single deep-vacuum method. The industry has therefore adopted another measurement called the *micron*. The micron is a unit of linear measure equal to 1/25,400th of an inch and is based on measurement above total absolute pressure, as opposed to gauge pressure, which can be affected by atmospheric pressure changes.

High-vacuum Indicators

To measure high vacuums, the industry developed electronic instruments. In general, these are heat-sensing devices, in that the sensing element, which is mechanically connected to the system being evacuated, generates heat. The rate at which heat is carried off changes as the surrounding gases and vapors are removed. Thus, the output of the sensing element (either thermocouple or thermistor) changes as the heat dissipation rate changes, and this change in output is indicated on a meter that is calibrated in microns of mercury. The degree of accuracy of these instruments is approximately 10 microns, thereby approaching a perfect vacuum.

Deep-vacuum Method of Evacuation

The single deep-vacuum method is the most positive method of assuring a system free of air and water. It takes slightly longer, but the results are far better.

Select a vacuum pump capable of pulling at least 500 microns and a reliable electronic vacuum indicator. The procedure is illustrated in Figure 4.14:

1. Install gauge manifold as already described.
2. Connect center hose to vacuum manifold assembly. This is simply a three-

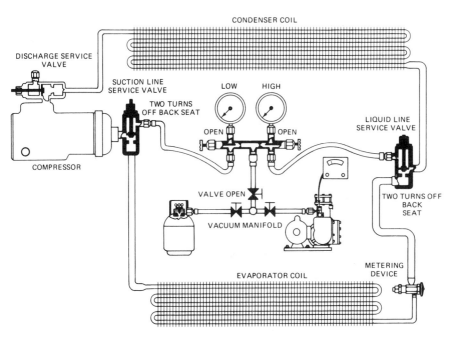

Figure 4.14   Deep vacuum evacuation assembly (Courtesy Carrier Corporation).

valve connection, allowing you to attach the vacuum pump and vacuum indicator, and a cylinder of refrigerant, each with a shut-off valve.

3. Open the valves to the pump and indicator. Close the refrigerant valve. Follow pump manufacturer's instructions for pump suction line size, oil, indicator location, and calibration.

4. Open (wide) both valves on gauge manifold and mid-seat both equipment service valves.

5. Start vacuum pump and evacuate system until a vacuum of at least 500 microns is achieved.

6. Close pump valve and isolate the system. Stop the pump for 5 minutes and observe the vacuum indicator to see if the system has actually reached 500 microns and is holding. If system fails to hold, check all connections for tight fit and repeat evacuation until system does hold.

7. Close valve to indicator.

8. Open valve to refrigerant cylinder and raise the pressure to at least 10 psig or charge system to proper level.

9. Disconnect pump and indicator.

Triple Evacuation

The triple-evacuation method requires no specialized high-vacuum equipment. However, this method should not be used if liquid water is suspected to be in the system. An evacuation pump of sufficient capacity to pull 28 in. of mercury vacuum will be needed. Good quality service gauges are important.

This method of evacuation is based on the principle of diluting the noncondensibles and moisture with clean, dry refrigerant vapor. This vapor is then removed from the system, carrying with it a portion of the entrained contaminants. As the procedure is repeated, the remaining contaminants are proportionately reduced until the system is contaminant-free. Figure 4.15 illustrates the assembly procedure:

1. Install gauge manifold as already described.
2. Connect center hose to vacuum manifold valves.
3. Connect pump and refrigerant cylinder to manifold valves. Purge lines with refrigerant.
4. Close refrigerant cylinder valve and open pump valve.
5. Open (wide) both valves on gauge manifold and mid-seat both service valves.
6. Start evacuation pump and evacuate system until a 28-in. mercury vacuum is reached on compound gauge. Let pump operate for 15 minutes at this level.
7. Close pump valve and stop pump.
8. Open refrigerant valve. Allow pressure to rise to 2 psig, then close refrigerant valve. Allow refrigerant to diffuse through system and absorb moisture for 5 minutes before next evacuation.

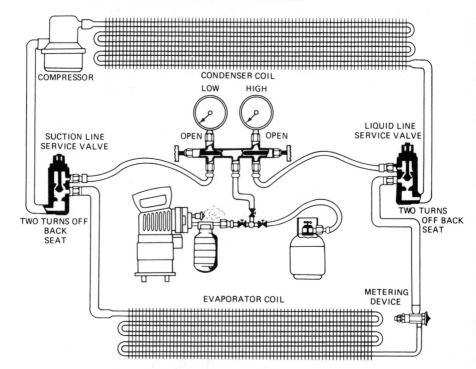

Figure 4.15   Triple evacuation assembly (Courtesy Carrier Corporation).

9. Close refrigerant valve. Open pump valve and repeat evacuation steps to again reach 28-in. mercury vacuum and hold for 15 minutes with pump running.
10. Close pump valve and turn off pump. Open refrigerant valve and charge to 2 psig, again holding for 5 minutes.
11. Close refrigerant valve. Open pump valve. Start pump and evacuate again to 28-in. vacuum and hold 15 minutes.
12. Stop pump and break vacuum, this time charging system to 1,099 psig or to proper level.

4.6 CHARGING THE SYSTEM

The quantity of refrigerant to be added to the system for initial charge or recharging depends on the size of the equipment and the amount of refrigerant to be circulated.

In smaller residential heat pump systems, the system refrigerant charge is *critical* to ounces rather than whole pounds. In this case, a charging cylinder is recommended, as illustrated in Figure 4.16. Refrigerant from the refrigerant

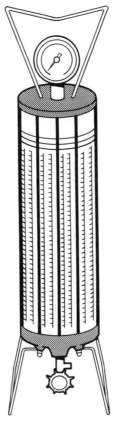

Figure 4.16   Charging cylinder.

drum is transferred to the charging cylinder. The charging cylinder has a scale that is visible to the operator so that precise measurement of the quantity of a specific refrigerant can be made to compensate for temperature and pressure conditions. These cylinders are accurate to 1/4 of an ounce. Optional electric heaters are available to speed the charging operation.

Where considerable installation and service work is involved, many contractors use a mobile evacuation and charging station. It contains a vacuum pump, charging cylinder, and service manifold and gauges. More elaborate models may also include a vacuum indicator and space for a refrigerant cylinder.

Charging Techniques

Refrigerant may be added in either the liquid or vapor form. Refrigerant is added in the vapor form when the unit is operating, through the suction valve. Refrigerant may be added in the liquid form when the unit is off and in an evacuated condition through the liquid-line service valve only.

Figure 4.17 illustrates the charging procedure for the vapor form (unit running). For simplicity we show only a refrigerant cylinder and assume the charge is weighed in during operation:

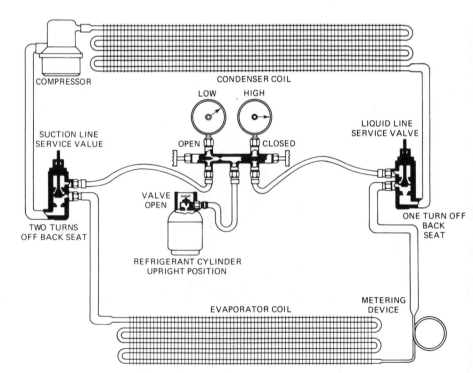

Figure 4.17   Vapor-form charging (Courtesy Carrier Corporation).

1. Install gauge manifold.
2. Attach refrigerant cylinder to center connection hose and open valve on low side of manifold.
3. Place cylinder in upright position.
4. Crack suction service valve two turns off back-seat.
5. Open valve on refrigerant cylinder and weigh in desired charge.
6. When correct charge has been added, close low-side manifold valve and valve on refrigerant cylinder.
7. Back-seat both suction and liquid-line service valves, remove hoses, and cap ports.

The charging procedure for the liquid form (unit not operating and evacuated) is shown in Figure 4.18.

1. Install gauge manifold.
2. Attach refrigerant cylinder. Invert cylinder, unless it is equipped with a liquid-vapor valve that permits liquid withdrawal in upright position.
3. Open both suction and liquid-service valves one turn off back-seat.
4. Open valve on high side of gauge manifold.
5. Open valve on refrigerant cylinder and add refrigerant.
6. After correct charge is introduced, close valve on high side of manifold

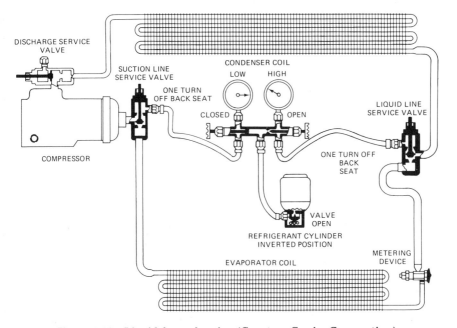

Figure 4.18   Liquid-form charging (Courtesy Carrier Corporation).

and close valve on refrigerant cylinder. Back-seat both suction and liquid-service valves.

7. Remove gauge manifold.

In both procedures the use of a charging cylinder would be recommended in smaller, critically charged systems where more accuracy is needed.

Checking the Charge

Checking the charge of a new installation or of an existing unit is another function of the service manifold gauges. For example, the following procedure would be used for an air-cooled unit:

1. Install gauge manifold.
2. Allow system to operate until pressure gauge readings stabilize (approximately 15 minutes).
3. While unit is operating, record the following information:
   a. High-pressure gauge reading.
   b. Dry bulb temperature of air entering condenser coil.
   c. Wet bulb temperature of air entering the evaporator coil. (This is performed with a wet-wick thermometer.)
4. A comparison of these measurements with the head pressure charging table supplied with the unit will indicate if the system is adequately charged and operating properly."[1]

4.7 ELECTRICAL METERS

Many of the problems in heat pumps can be traced to their electrical circuitry. For this reason, the heat pump technician will need a variety of electrical meters to test the electrical system. The following electrical meters should be considered standard in the heat pump technician's toolbox:

- Voltmeter: used to check voltages.
- Ammeter: used to check electrical current.
- Ohmmeter: used to check electrical resistance and continuity in electrical circuits; also used to check for shorts.

In lieu of these electrical meters, the technician may opt instead for a combination-type meter, such as the volt-ohm-ammeter, which

---

[1] Air-Conditioning and Refrigeration Institute, *Refrigeration and Air-Conditioning*, © 1979, pp. 28–30, 102–15. Reprinted by permission of Prentice-Hall, Inc., Englewood Cliffs, N.J.

allows the user to test for all three electrical properties (voltage, current, and resistance) using a single meter. However, these types of combination instruments, although handy, do not usually have as wide a range as their individual counterparts.

In some situations, the technician may be called upon to check out electrical circuits with low voltage and current measurements. Here, a milliammeter or millivoltmeter may be required, both of which are calibrated to record much smaller values than standard voltmeters and ammeters.

Regardless of the type of electrical meters that the heat pump technician decides to use, it is essential that he or she be tested on these instruments by someone experienced in electrical work before being allowed to use them in the field. The use of electrical meters does not require an extensive electronics background, but safety is of utmost importance, as is a basic understanding of electrical circuits.

Five

Installation and Start-up: Water-source Heat Pumps

Chapters 1–4 have covered the principles of heat pump operation and discussed the internal components. In this chapter, we will discuss the steps involved in the installation of a water-source heat pump, including a discussion of how to determine the method in which water will be supplied.

Of course, the first step in any heat pump installation (air or water source) will be to determine the cooling and heating needs of the residence in order to select the properly sized heat pump. A series of load calculations are performed to determine heat loss and heat gain, taking into consideration the size of the residence, type of insulation, number, type, and location of windows, location, climate, number of occupants, and other factors. Losses and gains are converted to BTUs (British Thermal Units), and these figures are used to determine the size and type of heat pump that will best meet the needs of a particular structure.

Many air conditioning and heat pump manufacturers provide their own forms for use in performing load calculations, and although similar procedures are used industry-wide, the actual manner in which load calculations are performed can vary greatly. However, ASHRAE, the American Society of Heating, Refrigerating, and Air-Conditioning Engineers, publishes a manual, *load calculation, Manual J*, that is the result of an overall industry-wide study of residential heat gains and calculation methods. Their procedures have become widely accepted, and many installers prefer them. Because of the abundance of material available from ASHRAE and other sources regarding the manner in which load calculations are performed, they will not be covered in this book. See Appendix A if you wish to obtain a copy of *Manual J* from ASHRAE, or you may wish to obtain load calculation forms from the manufacturer of the heat pump you are planning to install.

5.1 DESIGN CONSIDERATIONS: GROUND-WATER HEAT PUMP

After determining the size heat pump that will be required to meet the needs of the structure, the next step is to decide the manner in which water will be supplied to the heat pump. As mentioned in Chapter 2, there are two methods of delivering water to a residential water-source heat pump. Water may be supplied in the form of a well, which will require the services of a certified well driller; or in the form of a closed loop of buried piping that is filled with water, which will require the services of a certified contractor.

Generally, when a water-source heat pump has been selected for an installation, the normal order of progression is to first consider the use of well-supplied water, particularly if a well already exists. It is usually only after well-supplied water has been disqualified that a closed-loop system is considered, since a closed-loop system is usually more costly to install. If both types of systems can be used in a particular installation, the final decision will usually be left to the homeowner. Qualifying an installation for a ground-water heat pump system involves a number of factors, each of which is discussed here.

Type and Condition of an Existing Well

If a well already exists, the first step will be to determine its type and condition. There are three basic types of wells, although only one type is really suitable for use with a ground-water heat pump. These are the *hand-dug well*, which is usually a fairly shallow well (usually less than 50 feet) that is large in diameter and is lined with brick or stone; the *tile well*, which is also somewhat shallow (usually less than 100 feet deep), of large diameter and lined with porous tiles; and the *drilled well*, which is a small diameter hole that is bored with drilling equipment. Dug wells and tile wells are utilized in areas where abundant supplies of ground water exist at shallow depths, while the drilled well can be drilled to any depth to locate ground water. Drilled wells are considered to be the only type of well that are able to satisfactorily meet the water volume requirements of a ground-water heat pump.

If the existing well is of the drilled type, the installer should consult with a certified well driller to determine whether it is able to supply the requirements of the heat pump and continue to meet domestic water requirements as well. This usually involves performing a pumping test, in which a portable pump is set up a distance from the well and water is actually pumped at a specific flow rate over a period of time (ranging from 4 to 24 hours, depending on what is known about the well's past history and flow rate). During the test, water levels are checked periodically to determine whether the well's water-recovery rate is satisfactory. Both pumping capacity and water-recovery rate tests are extremely crucial. Should they be performed incorrectly or for an insufficient length of time, the integrity of the entire system, once installed, cannot be assured. A well must be able to provide the proper volume, and it must also be able to replenish itself. Always obtain the services of a reputable, certified well driller to perform these tests.

If an existing well is able to meet the combined needs of the home and the heat pump and has an adequate water-recovery rate, the first requirement of a ground-water heat pump has been met. This type of system is often referred to as a *combined system*. If this type of system is installed, a decision will need to be made whether to provide enough volume for both domestic and heating uses simultaneously or for the greater of the two uses. This is often up to the homeowner, who may want the additional comfort and convenience of adequate supply for all needs simultaneously, although such a system will require a larger pump and possibly the addition of a precharged pressure tank.

Pressure tanks, which are usually installed indoors near the heat pump unit, are used to provide compressed air that acts upon the water in the tank to force it to the intended point of use. A pressure tank contains a bladder or diaphragm (Figure 5.1) that separates the air from the water. The tank is usually precharged with air at the factory, and the tank is equipped with an air pressure valve to allow the installer to make adjustments if necessary. Because the water and air are permanently separated in the tank, only one water connection, which serves as both inlet and outlet, is required.

Assuming the pressure tank is filled with water, when water is demanded (for the heat pump or the home), pressure in the air chamber forces water into the system until the air chamber encompasses most of the interior of the tank and the pressure is reduced. When the pump cut-in pressure is reached, the pump starts automatically and begins to fill the tank, reducing the size of the air chamber and compressing the air. When the pump cut-off pressure is reached, the pump stops. In effect, then, the pressure tank actually replaces the well pump for brief periods of time, effectively reducing the number of times the pump must start and stop, and thus prolonging its useful life.

The use of a pressure tank requires the use of a larger pump, because the pump must be able to fill the pressure tank with water rapidly

DIAPHRAGM   BLADDER

AIR

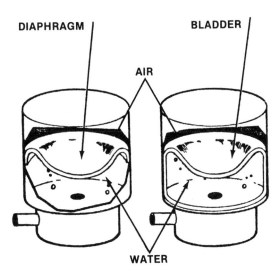

WATER

Figure 5.1 Pressure-tank diaphragm (left) and bladder (right) (Courtesy National Well Water Association).

on demand. Usually, a pump and pressure tank are selected to complement each other. In other words, the size of the tank is considered when selecting the size of the pump, and vice versa. Generally, however, the larger the tank, the better, keeping in mind cost and space considerations.

As an alternative to the additional cost of a pressure tank and a larger well pump to satisfy simultaneous needs, the installation of a reverse-acting pressure switch may be considered. A reverse acting pressure switch is installed in the water line between the pressure tank and the ground-water heat pump and serves to shut off the heat pump when water pressure drops below a preset minimum operating pressure. Here, the combined needs of the heat pump and the home would be considered when presetting a minimum operating pressure in such a way that the heat pump would only be shut down during peak domestic water usage. Since peak domestic water usage occurs for only a few minutes during an average day, actual shutdown time should be minimal and should not pose a major inconvenience. This is an extremely cost effective means of meeting both heat pump and domestic needs in a combined system.

Water Quality

If an existing well is able to supply both the heat pump and domestic needs, the next step is to consider the quality of the water. If the well is already being used to supply domestic water, it is logical to assume that the quality is satisfactory. However, since all well water contains some impurities, the water should be tested, preferably at a lab, to assure that high concentrations of iron, calcium, magnesium, salts, and other dissolved solids are not present, because such concentrations can result in degraded performance as a result of corrosion and scaling of the water coil. Again, this will be the responsibility of the well driller.

The water coil of a water-source heat pump is the component most likely to be affected by impurities in the water supply. Early water coils were constructed of copper, but newer models now utilize a cupro-nickel alloy, which is much more resistant to corrosion and scaling than copper. With the recent introduction of stainless steel water coils, corrosion and scaling as a result of poor water quality should be reduced significantly. Also, many heat pump manufacturers now design their equipment to reduce the effects of poor quality water, and often provide optional features for applications that require special considerations. With these improvements, it is likely that only in extreme cases will an existing well that is currently used to supply domestic water be disqualified. This usually occurs when concentrations of hydrogen sulfide, a highly corrosive agent, are detected in the water supply. Hydrogen sulfide is the most common corrosive agent, and its presence, even in very small concentrations, can result in disqualification of the well, since neither copper nor cupro-nickel show acceptable resistance to its effects.

Testing the water quality of an existing well will be more crucial if the well is not used to supply domestic water, since it is possible that the well has already been disqualified for home use due to poor quality. It should be understood that the water to be supplied for a heat pump does not have to be of drinking quality, so an existing well that cannot supply domestic needs due to poor quality can certainly meet the needs of a water-source heat pump. Here, however, the same lab and pumping tests should be performed, keeping in mind that we are dealing strictly with the needs of the water-source heat pump rather than the combined needs of the home and the heat pump. Again, though, concentrations of corrosive agents can result in disqualification of the well, although drinking quality will not be a factor.

To prevent chemical incrustation, every new, modified, or reconditioned water well, including all pumping equipment, should be disinfected by the well driller before being placed into service. This is normally performed with a chlorine solution made up of household bleach and water. If, however, bacterial infestation is found to be present, this may be cause for disqualification. Treating water for bacterial infestation is costly and not applicable in residential installations. Also, even if treatment were not so expensive, most local codes prohibit such treatment, since the water, after traveling through the heat pump, is returned to the ground and may contaminate the aquifer.

Return Well Considerations

If an existing well can meet the volume and quality requirements of a water-source heat pump, the next step is to consider the manner in which the discharged water will be dealt with.

As discussed in Chapter 2, the preferred method of discharge is to return the water to the supply aquifer if possible. The return well is drilled to the same aquifer at the same depth as the supply well, with the return well placed far enough away to prevent any overlapping thermal effect. Again, the services of a certified well driller should be obtained to assure that the system is in conformance with existing codes, because local codes vary widely regarding the use of return wells. Many states have no regulations regarding the disposal of return water from water-source heat pumps, but some states have enacted highly restrictive regulations.

The prime concerns of lawmakers are the temperature difference that occurs after water has traveled through the heat pump and the possible introduction of foreign substances into the ground water. It is these concerns that have prompted a few states to actually prohibit the use of a domestic supply well with a water-source heat pump. Also, some local authorities have expressed the fear that since the water-source heat pump can require up to 10,000 gallons per day in extremely cold weather, ground water supplies may be depleted if ground-water heat pump installations continue to increase. This, however, is usually not the case, since most installations simply return the water to the aquifer from which it was obtained, although some return the water to a different aquifer.

In urban areas, management of the heat balance in an aquifer may have to be coordinated to assure that thermal interference does not take place, since efficiency can be significantly reduced by overlapping heat exchange. Here, the well driller should take into consideration any other nearby water-source heat pumps utilizing well water, since random installation of numerous water-source heat pumps can lead to thermal interference through improper well spacing. It will be the responsibility of the well driller to assure that the installation is in conformance with all existing codes and regulations.

New Supply and Return Wells

If an existing well cannot satisfy the requirements of the water-source heat pump, a new supply well will have to be drilled along with a return well. It will be the responsibility of a certified well driller to satisfy the volume and quality requirements, install the pumping equipment, and provide for delivery of the supply water from the supply well to the residence, and discharge of return water to the return well. He or she will be responsible for selecting the appropriate piping sizes and materials based on the flow needed by the heat pump and a number of other factors. The well driller will also be responsible for installing an appropriate water filter at some point before entry to the heat pump to ensure that any solids are not introduced into the system. Other accessories, such as shut-off valves at various points in the system, will be installed by the well driller and will allow for periodic flushing and/or repairs that require isolation of sections of the water piping system. Other piping considerations that are normally the responsibility of the heat pump installer, but may also be performed by the well driller, will be discussed later in this chapter.

5.2 DESIGN CONSIDERATIONS: CLOSED-LOOP HEAT PUMP

If, after consulting with a certified well driller, it is found that there is an insufficient or poor-quality water supply, a closed-loop system should be considered.

A closed-loop system utilizes heat transfer between the soil and the water within the loop piping. It is important that the system be designed to offset water temperature changes that occur during heat pump operation with the best transfer ability of the soil surrounding the loop piping. In other words, the loop must be designed so that it is able to adequately absorb heat during the cooling cycle and release heat during the heating cycle, in such a way that ground temperatures are not raised or lowered excessively. This usually involves providing a large enough loop of piping to allow for balanced rises and falls in water temperatures. This is the major design consideration of a closed-loop system.

It should be understood that the installation of a closed-loop system, regardless of type, will require the services of an experienced contractor. Many of the major manufacturers of water-source heat pumps provide installers with the names of experienced contractors in their area. However, because closed loops are relatively new and because of the extensive work being performed at Oklahoma State University (OSU), it is highly recommended that an installer contact the university directly if a closed-loop system is being considered. OSU provides a special extension service that may be contacted by phone to aid in the design of a closed-loop system. They are the best source for information on closed-loop designs and can provide relevant soil and weather information that will directly relate to the manner in which a system is designed for a particular area of the country. See Appendix A for their address and phone number.

The primary design consideration in a closed-loop system is whether to use a vertical or horizontal loop. A vertical loop consists of one or more vertical boreholes (Figure 5.2), in which plastic pipe is inserted. Obviously, a vertical loop requires considerably less land area and is usually selected when space constraints will not allow for a horizontal loop. Also, because vertical installations can go up to 200 feet or deeper into the Earth, warmer soil conditions will exist, which translates into a decreased amount of overall piping in the system. However, both vertical and horizontal loops require the addition of an antifreeze solution into the system, although the amount will be less for a vertical loop. A closed loop also requires the installation of a pumping kit that is usually offered as an optional accessory by the heat pump's manufacturer. The installation of the closed-loop pump kit is normally the responsibility of the heat pump installer rather than the contractor. This will be covered later in the chapter.

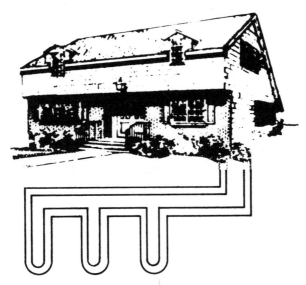

Figure 5.2   Vertical loop (Courtesy National Well Water Association).

A horizontal loop (Figure 5.3) is similar to a vertical loop, except the piping is installed in the bottom of a trench that is dug to a minimum depth of 6 feet with a trenching machine. A horizontal loop requires more space than a vertical loop, and the amount of piping is based upon soil characteristics, temperatures, and the capacity of the heat pump. OSU should be consulted, because they have worked out standard calculations that can be used to determine the type and size loop for a particular installation. OSU can also refer an installer to an experienced contractor in a particular area of the country or provide instructional

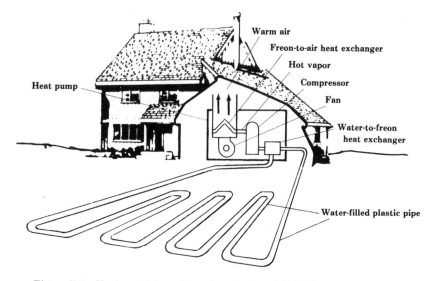

Figure 5.3   Horizontal loop (Courtesy National Well Water Association).

material to compensate for a contractor who has no experience in installing closed-loop systems.

Generally, the contractor will be responsible for all trenching or drilling and for installing the loop piping in the ground. Flushing and testing the loop before connecting loop piping to the pumping module and charging the system with antifreeze may also be the contractor's responsibility, but the heat pump installer should be familiar with these procedures, because they may need to be performed at a later time if the system needs repairs. Flushing and testing a closed-loop system will be covered later in this chapter.

5.3 INSTALLING THE HEAT PUMP

Location

The location of the heat pump should be carefully considered before installation to ensure adequate clearance to any service panels on the unit. Generally, a space of 18 inches or more should be provided for clearance, although local codes should be adhered to where applicable. Sufficient space should also be provided to make all water, electrical, and duct connections. If the unit is being installed in a confined space, care should be taken to ensure that adequate provision is made for return air to freely enter the space. The manufacturer's specifications should be consulted for the minimum ambient temperatures under which the unit can satisfactorily operate, because some units cannot be installed in areas where ambient temperatures fall below freezing.

Ductwork

If the heat pump is being installed as a replacement for an existing heating system or as an add-on, all existing ductwork must be checked to ensure that it meets the air-flow requirements of the heat pump being installed. Figure 5.4 shows a residential duct-sizing chart for one manufacturer's heat pumps. Such charts are normally included with the heat pump unit, and should be consulted to determine if an existing ductwork system can be used or whether it will be necessary to make modifications. Duct-sizing charts are referred to for new homes as well. Charts such as this can be obtained from ASHRAE (See Appendix A).

If air leaks are discovered, they may be wrapped with duct wrap.

RESIDENTIAL DUCT SIZING CHART

Standards: 900 FPM Main Duct
600 FPM Branch

Design duct size at .085 — .1 friction per 100 feet

	ACCEPTABLE BRANCH DUCT SIZES		ACCEPTABLE MAIN OR TRUNK DUCT SIZES	
CFM	Round	Rectangular	Round	Rectangular
50	4″	4 x 4		
75	5″	4 x 5, 4 x 6		
100	6″	4 x 8, 5 x 6		
125	6″	4 x 8, 5 x 6, 6 x 6		Metal duct in unconditioned space must have liner or be wrapped with blanket insulation of 1″ thickness. Fiberglass board must be jointed by fab and mastic or stapled with heat-applied aluminum tape. Grey-duct tape or plain-foil tape are not acceptable.
150	7″	4 x 10, 5 x 8, 6 x 6		
175	7″	5 x 10, 6 x 8, 4 x 13, 7 x 7		
200	8″	5 x 10, 6 x 8, 4 x 14, 7 x 7		
225	8″	5 x 12, 7 x 8, 6 x 10		
250	9″	6 x 10, 8 x 8, 4 x 16		
275	9″	4 x 20, 8 x 8, 7 x 10, 5 x 15, 6 x 12		
300	10″	6 x 14, 8 x 10, 7 x 12		
350	10″	5 x 20, 6 x 16, 9 x 10		
400	12″	6 x 18, 10 x 10, 9 x 12	10″	4 x 20, 7 x 10, 6 x 12, 8 x 9
450	12″	6 x 20, 8 x 14, 9 x 12, 10 x 11	10″	5 x 20, 6 x 16, 9 x 10, 8 x 12
500			10″	10 x 10, 6 x 18, 8 x 12, 7 x 14
600			12″	6 x 20, 7 x 18, 8 x 16, 10 x 12
800			12″	8 x 18, 9 x 15, 10 x 14, 12 x 12
1000			14″	10 x 18, 12 x 14, 8 x 24
1200			16″	10 x 20, 12 x 18, 14 x 15
1400			16″	10 x 25, 12 x 20, 14 x 18, 15 x 16
1600			18″	10 x 30, 15 x 18, 14 x 20
1800			20″	10 x 35, 15 x 20, 16 x 19, 12 x 30, 14 x 25
2000			20″	10 x 40, 12 x 30, 15 x 25, 18 x 20

Grilles and registers shall be sized according to manufacturer's performance data capable of handling the CFM of the duct at a throw based on room dimensions. Return air registers should be selected to provide for 450 FPM face velocity.

Figure 5.4 Residential duct-sizing chart (Courtesy Command-Aire Corporation; Waco, Texas).

If the heat pump is installed in an unconditioned space, any ductwork within that space must be insulated to prevent significant heat loss.

Mounting the Heat Pump

Water-source heat pumps may be of either horizontal or vertical configuration. Horizontal units, if used at all in residential applications, are often suspended at ceiling level by means of an optional mounting kit offered by the manufacturer of the heat pump. Such a kit usually consists of four threaded rods that are anchored into the ceiling and then attached to heavy brackets that support the unit, as shown in Figure 5.5. A horizontally installed heat pump will also require a drain pan that is placed under the entire length of the unit. Horizontal units can also be mounted on a floor or possibly on a shelf, again with an appropriate drain pan.

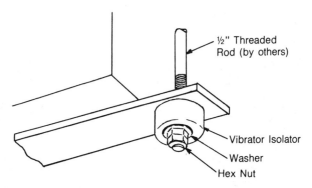

Figure 5.5 Horizontal mounting kits (Courtesy Friedrich Climate Master, Inc.).

Vertical units, which are more common in residential applications, are most often mounted on a floor or any reasonably level and flat surface that is of adequate strength to support the weight of the unit. They may be set on a concrete floor or on a rigid platform. The platform must be constructed of steel or wood, and must support the entire bottom of the unit. The unit is then mounted on a pad of rubber-like material (as a sound insulator) that is slightly larger than the base to provide isolation between the unit and the floor. It is not necessary to anchor a unit to the floor.

Electrical Connections

All installations should be completed in accordance with local electrical code. Units should be connected to supply boxes with copper wire only. Aluminum wire is to be avoided.

The power supply should be 60 Hertz, single or three phase, sized for the current and voltage required by the unit being installed. This information is available from wiring diagrams provided with the unit, which illustrate all internal wiring. The power supply to the unit should be armored cable, or it should be run through flexible conduit. It should be run from a separate disconnect that is installed within 3 feet of the unit. Conduit or armored cable should be positively grounded to the unit.

Location of the Thermostat

The location of the thermostat should be in the main flow of air back to the unit or return registers. The thermostat should never be placed on an outside wall or in a kitchen. It should be mounted approximately 5 feet from the floor, and only approved thermostats for the unit being installed should be used to assure compatibility. All thermostat wiring should be connected in strict accordance with wiring diagrams supplied with the unit being installed.

Water Piping Accessories and Considerations

Regardless of the manner in which water is supplied (well or loop), a number of accessory items can make the entire system more efficient and, more importantly, easier to maintain and repair. The need for such accessory items will be determined by the installer, because he will normally be the one who must make periodic checks on the system during its useful life.

One accessory is a water filter or strainer, which is normally installed only when well water is being utilized. The filter is placed near the heat pump and serves to trap any solids that may be present in the water supply.

The installation of a solenoid valve at the discharge side of the heat pump is used to keep the water under pressure through the water coil. Gate valves and boiler drains on both the inlet and outlet sides of the heat pump will allow for cleaning of the water coil. A detailed diagram of heat pump connections and recommended piping arrangements is shown in Figure 5.6.

The recommended water pipe size, as mentioned earlier in this chapter, will depend on the gallons per minute (gpm) requirements of the unit being installed. The basic piping installation includes a ball valve in the water supply line prior to the unit, with an on/off flow control valve on the discharge line, as shown in Figure 5.7. A slow opening motorized water valve controlled by low voltage is recommended to open and close the line as the unit is called into operation

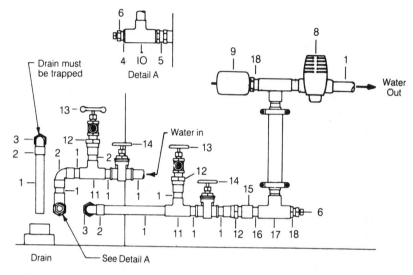

Figure 5.6 Recommended piping arrangements (Courtesy Command-Aire Corporation; Waco, Texas).

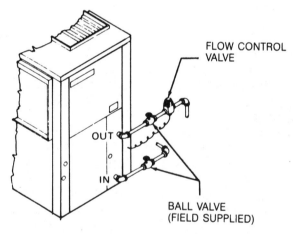

Figure 5.7 Basic piping installation (Courtesy Command-Aire Corporation; Waco, Texas).

by the thermostat. Solenoid valves will work, but tend to be noisy and may distort the temperature swing during the cooling cycle due to the extra load on the low-voltage system.

Pressure-activated water regulating valves are recommended in areas with water temperatures below 60°F. In such instances, a pair of pressure-activated water regulating valves are recommended to control water flow. One valve is reverse acting (low-pressure range), and opens on the fall of refrigerant suction pressure in the heating mode. The other valve is direct acting (high-pressure range), and opens on a rise of refrigerant discharge pressure in the cooling mode of operation (see Figure 5.8).

Figures 5.9 through 5.13 illustrate the recommended transition procedures for linking various materials used in water piping. It is recommended that PVC plastic pipe be avoided in any piping for an earth-coupled system, because PVC tends to crack under temperature stress,

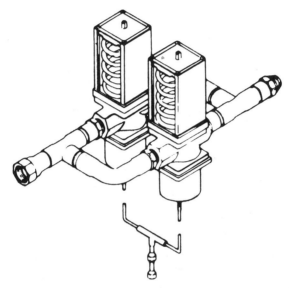

Figure 5.8 Water-regulating valve (Courtesy Command-Aire Corporation; Waco, Texas).

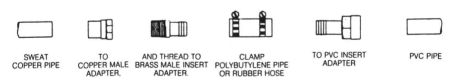

SWEAT COPPER PIPE	TO COPPER MALE ADAPTER.	AND THREAD TO BRASS MALE INSERT ADAPTER.	CLAMP POLYBUTYLENE PIPE OR RUBBER HOSE	TO PVC INSERT ADAPTER	PVC PIPE

Figure 5.9 PVC-to-polybutylene connections (Courtesy Command-Aire Corporation; Waco, Texas).

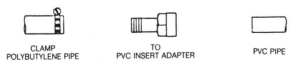

CLAMP POLYBUTYLENE PIPE	TO PVC INSERT ADAPTER	PVC PIPE

Figure 5.10 PVC-to-copper connections (Courtesy Command-Aire Corporation, Waco, Texas).

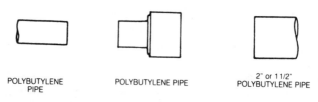

POLYBUTYLENE
PIPE

POLYBUTYLENE PIPE

2" or 1 1/2"
POLYBUTYLENE PIPE

Figure   5.11 Polybutylene-to-polybutylene   connections   (Courtesy
Command-Aire Corporation; Waco, Texas).

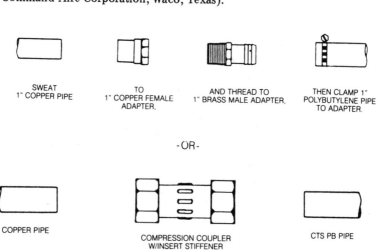

SWEAT
1" COPPER PIPE

TO
1" COPPER FEMALE
ADAPTER,

AND THREAD TO
1" BRASS MALE ADAPTER,

THEN CLAMP 1"
POLYBUTYLENE PIPE
TO ADAPTER.

-OR-

COPPER PIPE

COMPRESSION COUPLER
W/INSERT STIFFENER

CTS PB PIPE

Figure 5.12   Polybutylene-to-copper connections (Courtesy Command-
Aire Corporation; Waco, Texas).

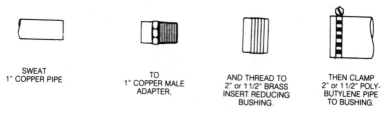

SWEAT
1" COPPER PIPE

TO
1" COPPER MALE
ADAPTER,

AND THREAD TO
2" or 1 1/2" BRASS
INSERT REDUCING
BUSHING.

THEN CLAMP
2" or 1 1/2" POLY-
BUTYLENE PIPE
TO BUSHING.

Figure 5.13   Reducing connections (Courtesy Command-Aire Corporation; Waco, Texas).

resisting the necessary expansion. Should the installer decide to use
materials other than those described in Figure 5.9 through 5.13, it is
recommended that the material manufacturer be consulted for detailed
and accurate instructions regarding connections and transitions to other
materials.

Condensate Drain Piping

When the heat pump is in the cooling mode, condensate will form on the air coil and drip into a drain pan (provided with the unit). A piping run should be installed to drain this condensate to a sanitary drain to prevent condensate overflow from damaging the heat pump. In most cases, the drain line should have a trap using 90° ells (Figure 5.14), so that the inside fan or blower suction does not pull the condensate back into the unit. A preformed trap is recommended. PVC is recommended and should be of proper gauge to prevent sweating. Local codes, where applicable, should be strictly adhered to.

Figure 5.14 P trap (Courtesy Command-Aire Corporation; Waco, Texas).

Flowmeters

A flowmeter provides a means of checking adequate flow of water to the heat pump. It is often permanently installed at the time of installation of the system. This can be done by mounting the flowmeter on the inlet side of the heat pump, or by means of a series of valves that are used to divert water to a separate receptacle, with the latter method being applicable only to ground-water heat pumps. Figure 5.15 shows a typical flowmeter installation.

Pete's Plugs

An alternative to installing a permanent flowmeter is the utilization of "Pete's Plugs" in the system to make accurate performance measurements. These plugs are installed on the inlet and outlet connections of the water coil and are used to determine both temperature and pressure differentials. The advantage to using Pete's Plugs is that the overall cost of the system is reduced. However, while flowmeters require only a thermometer capable of taking skin-temperature readings, systems utilizing Pete's Plugs require a probe-type thermometer, an adapter, and a pressure gauge.

A Pete's Plug is screwed into a tee in the water line. The plug contains a valve very similar to those used on basketballs. Temperatures are taken by forcing a 1/8″ probe of a dial thermometer through the neoprene disc. When the probe is removed, the neoprene seals itself.

Flowmeter Installation with Plastic or Polybutylene Pipe

Install the flowmeter in a vertical PLASTIC piping run.

DO NOT THREAD METAL FITTINGS INTO THE FLOWMETERS. Use only 1 inch PVC adapters.

If using coiled 1 inch polybutylene on either side of the flowmeter, use short pieces to avoid stressing the flowmeter.

Use Teflon or similar tape on the connections.

Tighten the PVC adapters only enough to keep the connections from leaking. Over-tightening may cause damage to the flowmeter. DO NOT hold the flowmeter body with pliers during the installation.

To avoid melting the polybutylene pipe, PVC adapters or flowmeter, solder any joints in the vicinity of the flowmeter first and allow them to cool before installing the plastic pipe, fittings or flowmeter.

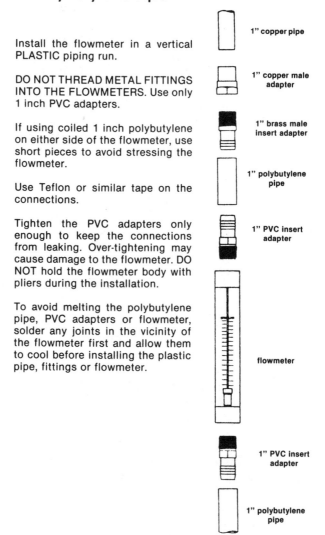

1" copper pipe

1" copper male adapter

1" brass male insert adapter

1" polybutylene pipe

1" PVC insert adapter

flowmeter

1" PVC insert adapter

1" polybutylene pipe

Figure 5.15 Flowmeter installation (Courtesy National Well Water Association).

5.4 CLOSED-LOOP PUMP KITS

In closed-loop installations, a pump kit is required to meet the pumping, air elimination, and fluid expansion requirements of the system. Most closed-loop heat pump manufacturers offer these kits as an optional accessory, and although each kit may be different, they all work in basically the same manner and perform the same basic functions.

A pump kit contains a low-wattage, loop-circulating pump, a manual-release air vent, air purger, thermal expansion tank with a fluid thermometer, a water-flow meter, pressure gauge, isolation valves and two service valve assemblies, and a low-temperature freezestat. They are usually sold in preassembled modules, making them fairly simple to install. Service and performance checks are aided by built-in flow, temperature, and pressure indicators, and the service valve assemblies provide flush-out, fill, and fluid-access ports, in addition to reducing from outside to inside pipe sizes.

Figure 5.16 shows a typical pump kit installation in conjunction with a water-source heat pump. Note that the pump kit is installed as close to the heat pump as is practical, with the flowmeter in a vertical position. Direct stud attachment, where applicable, is recommended to adequately support the pump kit.

The type and size of piping will be specified by the heat pump manufacturer, as will be the recommended fittings. PVC, however, is generally unsuitable because plastic to metal-threaded connections are difficult to avoid when it is used. Plastic to metal connections must be avoided, because the dissimilar rates of expansion of these two materials over loop temperature swings will inevitably cause leaks. Metal

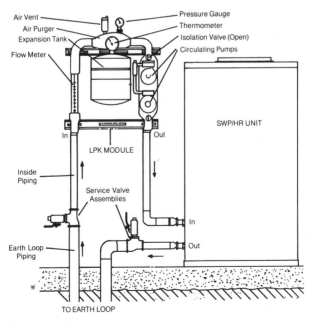

Figure 5.16 Pump kit installation (Courtesy Command-Aire Corporation; Waco, Texas).

insert adapters should be used for all transitions to metal connections, and plastic insert fittings should be used for all other connections. Also, all inside piping must be insulated to prevent sweating.

After the pump kit and heat pump have been installed, the entire system should be pressure tested and inspected for leaks before insulating over connections. This can be accomplished by pressurizing the system with the domestic water supply through a hose connected to one of the service valves.

Flushing and Testing a Closed Loop

It is imperative that a closed loop be power flushed prior to the final fill and antifreeze charging, in order to eliminate all of the air and possible debris in the system. Any air in a closed loop can accumulate and block flow, which will severely impede system performance.

Referring to Figure 5.17, the following flushing procedure should be observed:

1. Connect a flushing pump (minimum 1 1/2 hp capable of 35 psi) and a storage drum to service valve assemblies with the direction of flow as indicated in Figure 5.17. All connecting hoses must be 1 1/4″ inner diameter or larger.

2. Close one isolation valve on the pump kit to force all flushing flow through the loop.

3. Fill the storage drum with clean water, open both service valves, and start the flushing pump. Depending on the amount of water required to fill the loop, additional water may need to be added to the storage drum. The drum need not be full, but the flushing

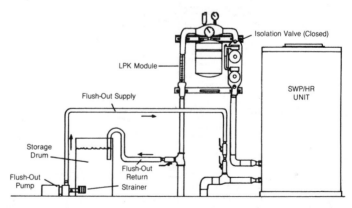

Figure 5.17 Flushing a closed loop (Courtesy Command-Aire Corporation; Waco, Texas).

pump suction line must never draw air or sediment from the bottom of the drum (a suction line is recommended). Keep the flushing return line under the water level in the drum to prevent foaming and air entrapment.

4. For parallel loops only: If the header is still accessible (highly recommended), pinch off flow paths one at a time and ensure that the flushing flow rate drops by observing the stream out of the flushing return line. If a particular path produces no change in flow when pinched off, it is not flowing properly; the cause of blockage should be determined and corrected. Possible causes of blockage are air locks, kinks, foreign materials in the lines, or saddle fusion taps not drilled out.

5. Flush the system until no air bubbles are observed coming out of the flushing return line into the storage drum. This may take 10–20 minutes.

6. Close the service valves, stop the flushing pump, and reverse the flushing direction through the loop, as shown in Figure 5.18. Be certain that the flushing supply line remains completely full of water so that no air is injected into the loop on restart. A suction line foot valve or an additional ball valve at the end of the supply line can be used to prevent air from entering the line during changeovers.

7. Start the flushing pump, open service valves, and flush the system as previously described until all air is eliminated in this flow direction.

8. While still flushing, open the isolation valve on the pump kit to force a flush-out of the inside piping and the heat pump. This flushing flow should be visible in the flowmeter. It is advisable to

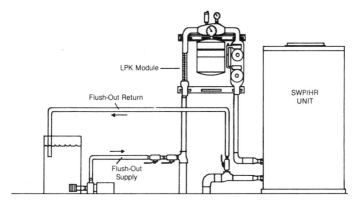

Figure 5.18 Reverse flushing a closed loop (Courtesy Command-Aire Corporation; Waco, Texas).

apply power to the pump kit's circulation pump during this procedure to achieve maximum flow through the inside system. Continue this until all air is eliminated.

9. Close the service valves and stop the flushing pump.

Charging a Closed-loop System with Antifreeze

After the power flush has been completed and the system has been checked for leaks, the system is ready for antifreeze charging. The two most common solutions used in closed loops are calcium chloride and proprolene glycol, with calcium chloride being less expensive and more widely used.

The amount of antifreeze used in a system will depend on the total volume of water contained in both the outside and inside piping of the loop system. The heat pump manufacturer will usually provide a chart, such as that shown in Figure 5.19, to enable the installer to determine the proper amount of antifreeze charging for a particular installation based on the type, size, and amount of pipe.

PIPE	SIZE	Volume (Gal./100' Pipe)
Copper	1"	4.1
Copper	1¼"	6.4
Copper	1½"	9.2
Polybutylene	1" IPS	3.7
Polybutylene	¾" IPS	2.8
Polybutylene	1" IPS	4.5
Polybutylene	1¼" IPS	7.8
Polybutylene	1½" IPS	11.5
Polybutylene	2" IPS	18.0

Add five (5) gallons to allow for the heat pump and LPK.

Figure 5.19 Approximate fluid volumes of pipe (Courtesy Command-Aire Corporation; Waco, Texas).

The following procedure should be observed for liquid antifreeze charging of a closed loop:

1. Hook up a flushing pump and storage drum, as shown in Figure 5.20. Close the pump kit's isolation valve so that the antifreeze cannot bypass through the pump kit to the drain line. Ensure that the flushing supply line (pump discharge line) is completely full of fluid so that no air is injected into the system.

2. Empty the water from the storage drum and fill it with the proper

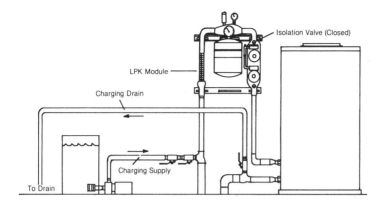

Figure 5.20 Antifreeze charging a closed loop (Courtesy Command-Aire Corporation; Waco, Texas).

amount of antifreeze. If the drum is not large enough, fluid may be added while charging.

3. Start the flushing pump and open both service valves. Water being displaced by the antifreeze will flow out of the drain line. Continue until all the antifreeze is pumped into the system.

4. Close the drain line service valve to build pressure in the loop system while observing the pump kit's pressure gauge. When the proper operating pressure is reached, close the supply line service valve and stop the flushing pump.

5. Open the pump kit's isolation valve and start the loop circulating pump to check for proper operation and flow rates.

6. Remove the flushing/filling apparatus and install plugs in the service valves.

Checking the Flow Rate in a Closed-loop System

It is imperative that the system's flow rate be checked at this point. To test the flow rate, open the pump kit's isolation valves and start the heat pump. The manner in which the flow rate is checked will depend on the type of installation (flowmeter, Pete's Plugs, and so on). Observe each manufacturer's procedure regarding their method of measuring flow rate.

If system flow is low or absent, remove the indicator plug from the center of the circulating pump's housing and ensure that the shaft is spinning. A small screwdriver may be used to rotate the shaft manually if it is stuck. Other causes for inadequate flow are air locks, blocked or kinked piping, no power, pipe shavings in the circulating pump or heat pump, or improperly sized loop piping. Be sure to loosen the cap on the air vent during this procedure.

After the system circulates for a while, it may need to be pressurized again, because the air vent releases entrapped air (on improperly flushed loops). Pressurizing the system can be easily accomplished with a hose connecting from the domestic supply into one of the service valves. Purge all air from the hose before connection.

NOTE: The system pressure will drop as the plastic earth loop pipe relaxes, and will also fluctuate with changes in the fluid temperature. Also, if the system is opened for any reason at any time, the system must be reflushed using the procedure described above.

5.5 START-UP

Checklist

After the installation is complete and the system has been completely flushed, filled, and purged of air, the next step is to start the heat pump and compare its performance to charts and tables supplied with the unit.

Before actually starting the heat pump, however, a number of checks should be made to ensure that all installation procedures have been performed. Different steps may be required than those I will describe, depending on the unit and the manner in which water is supplied. However, the following steps are somewhat standard and should be observed:

1. High voltage is correct.
2. Fuses, breakers, and wire size are correct.
3. Low-voltage wiring is complete.
4. Water system has been cleaned and flushed.
5. Air is purged (closed-loop system only)
6. Water control/isolation valves are open.
7. Condensate line is open, trapped, and correctly pitched.
8. Blower rotates freely.
9. Blower speed is correct.
10. Air filter is clean and in position.
11. Shipping restraints are removed from the compressor.
12. Service/access panel is in place.
13. Thermostat is in the "OFF" position.

If all of the above checks out, the unit is ready for start-up.

Start-up Procedure

Each manufacturer's specific start-up procedures should be observed, particularly those that apply to thermostat operation, because different thermostats may have different types of settings that should be observed. The following list, however, can be considered a standard start-up procedure:

1. Turn the thermostat fan position to "ON." The blower should now operate.
2. Balance air flow at registers, referring to manufacturer's specifications provided with the heat pump unit.
3. Set the thermostat to the highest temperature.
4. Set the thermostat switch to the "COOL" position. The compressor should *not* come on.
5. Slowly reduce the thermostat setting until both the compressor and water flow are activated. Verify that the compressor is on and that the water flow rate is correct.
6. Check the temperature of both the supply and discharge water, comparing them with specifications provided with the heat pump unit.
7. Check refrigerant pressures, comparing them with specifications provided with the heat pump unit.
8. Check for an air temperature drop of $15°F$ to $20°F$ across the air coil.
9. Turn the thermostat switch to the "OFF" position. A hissing sound indicates proper functioning of the reversing valve.
10. Leave the unit off for approximately 5 minutes to allow pressures to equalize.
11. Turn the thermostat to the lowest setting.
12. Set the thermostat switch to the "HEAT" position.
13. Turn the thermostat dial to higher temperatures until both the compressor and water flow are activated.
14. Check refrigerant pressures, comparing them with specifications provided with the heat pump unit.
15. Check for an air temperature rise of $20°F$ to $35°F$ across the air coil.
16. Check that the amperage draw is correct.
17. Turn the thermostat to a higher setting to make certain that the auxiliary electric heat is activated (if installed), with the appropriate rise across the heater.

18. Confirm that the auxiliary heat comes on when the thermostat is in the "EMERGENCY HEAT" mode, if one is available on the thermostat. Most thermostats have such a setting.
19. Check for vibrations, noise, and water leaks.
20. Set the system to maintain the desired comfort level.
21. Instruct the owner/operator as to the correct thermostat settings and system operation in general.

During system start-up, if any of the above procedures indicate a problem, refer to Chapter 8 for troubleshooting procedures.

5.6 INSTALLATION: WATER-TO-WATER UNITS

In some installations, it is desirable to utilize a water source (well or loop) to heat domestic water in conjunction with standard heating and cooling. This section will describe how this is accomplished by different manufacturers.

Utilizing a heat pump to provide domestic hot water requires an additional coil at some point in the system to capture excess heat energy from the heat pump. Generally, water-source heat pump manufacturers offer optional equipment that allows for hot water heating in conjunction with their heat pump units; the methods utilized vary, as do the installation procedures. Command-Aire, for example, offers a circulating pump assembly that is attached directly to their heat pump, which comes equipped with an extra, built-in heat exchanger for hot water heating purposes. The pump assembly comes with the circulating pump and all necessary fittings and valves to allow for direct connection to the desuperheater of the heat pump and an existing hot water tank. Also supplied is a water temperature thermometer with well that is permanently installed in the system.

Friedrich Climate Master offers a hot water generator unit that is intended for retrofit on any existing refrigeration, air conditioning, or heat pump systems. It is a separate, packaged unit (containing the additional heat exchanger required) that may or may not be used in conjunction with a heat pump. When used with a heat pump, the hot water generator is connected to the heat pump by means of piping between the compressor and the reversing valve; the unit is installed as close as possible to both the heat pump and the existing hot water tank.

FHP Manufacturing offers hot water heating by means of their FHP Energymiser Hot Water Heat Recovery unit, which is installed in conjunction with their heat pump. It is also a separately packaged unit that contains the required additional heat exchanger, but the unit, when installed, becomes an integral part of the heat pump. The same principles are utilized to heat the water, although the installation and the manner in which hot water is obtained is different.

Because each manufacturer offers a different type of equipment and the installation of each unit varies, it would not be appropriate to detail each installation in this book. When installing a hot water heating unit in conjunction with a water-source heat pump, the individual manufacturer's installation and operating manuals should be consulted.

Six

Installation
and Start-up:
Air-source Heat
Pumps

Unlike the water-source heat pump, the installation of which may require coordination among two and possibly three contractors, the air-source heat pump can usually be installed by a single technician. Likewise, pre-installation steps and design considerations are significantly reduced as well, since the air-source heat pump installation is fairly straightforward, regardless of the structure. However, the technician must first perform a series of load calculations to determine the type and size of unit to be installed based upon the needs of the structure. It will be up to the technician to decide which load calculation forms to utilize, and the specific manner in which load calculations are performed (see Appendix A to obtain information from ASHRAE regarding load calculations).

This chapter will outline the general steps involved in the installation of an air-source heat pump, regardless of its packaging. Remember, an air-source heat pump can be a packaged system (single unit only), a split system, or an add-on split system. If the unit is packaged, it will be installed outdoors. The major difference between a packaged unit and the outdoor section of a split system or add-on split system will be size and weight, because all components are housed in a single package.

6.1 LOCATION AND MOUNTING OF THE OUTDOOR UNIT

Whether the outdoor unit is the entire unit or the outdoor section of a split system or add-on split system, the same guidelines will apply. There will be different factors to consider depending on whether the unit is being installed at ground level or on a rooftop. The greater majority of residential air-source heat pumps, however, are installed at ground level.

Ground-level Installation

1. Locate the unit away from overhanging roof lines that could allow water or ice to drop on or in front of the unit.

2. An attempt should be made to locate the unit where prevailing winds cannot blow directly on the coil. If this is not possible and the unit is being installed in an area of the country where ambient temperatures are quite low, a wind barrier may be constructed of at least the same height and width of the unit. Such a barrier should be placed approximately 2 feet away from the unit.

3. In areas where snowfall is likely, the unit should be mounted high enough to be above the average area snowfall to ensure proper drainage. Local codes and/or requirements, however, should be observed at all times. Elevating the unit can be accomplished by constructing a frame using suitable materials. If a support frame is utilized, care should be taken that drain holes in the unit base are not blocked in any way.

4. A sound-absorbing material may be installed under the unit to prevent the transmission of sound and/or vibration to the interior of the structure.

5. If the unit is mounted on a slab (concrete or other suitable material), the slab should be pitched slightly away from the building to allow for adequate drainage.

6. The unit should be installed with sufficient clearance for air entrance to the outdoor coil, for proper air discharge, and for service access. Refer to the installation manual for the unit being installed for minimum clearances. Generally, the location should provide a minimum of 2 feet from walls and shrubbery and a vertical clearance of 5 feet, as shown in Figure 6.1.

Rooftop Installation

1. All building codes should be consulted to ensure that proper weight distribution requirements are met.

2. The roof should be checked to ensure that unit support is adequate.

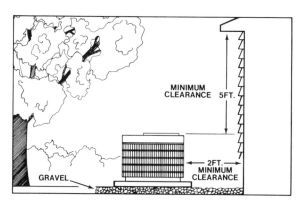

Figure 6.1 Minimum clearances.

3. The unit should be located on a solid, level platform that will allow minimum clearance between the rooftop and the platform.

6.2 LOCATION OF THE INDOOR UNIT

The location of the indoor unit of a split-system heat pump will depend upon a number of factors. Most indoor units are approved for installation in basements, utility rooms, closets, attics, or even crawl spaces, although some manufacturers may specify minimum ambients at which the unit can be expected to operate satisfactorily. Also, since some units are vertical and some are horizontal, the installation will vary depending upon configuration.

The following guidelines can be generally observed:

1. The indoor unit must be correctly matched to the outdoor unit. Units from different manufacturers are generally incompatible.
2. The indoor unit should be located as close to the outdoor unit as possible and positioned to avoid as many refrigerant piping bends as possible.
3. The unit should be located to observe minimum clearances specified for the unit being installed.
4. The unit should be centralized with the air-distribution system if possible.
5. The unit must be located to permit installation of a trapped condensate line to an open drain.
6. Units being installed vertically or horizontally can be set directly on a floor or platform, or they can be supported by metal or wooden beams. Units being installed can be suspended from above.

Oil Return Considerations

In a split system, it is mandatory that the satisfactory return of oil to the compressor is assured. If the units are installed near level to each other, the suction line should be pitched toward the outdoor unit. To prevent liquid from draining to the compressor during shutdown when the indoor unit is installed above the outdoor unit, the suction line should incorporate a rise above the level of the indoor unit by means of an inverted loop. When the indoor unit is installed 4 feet or more below the outdoor unit, the suction line must contain an oil trap at the base of the rise to insure proper oil return to the compressor. Additional traps may be required for higher rises. Figure 6.2 provides some examples of different types of installations.

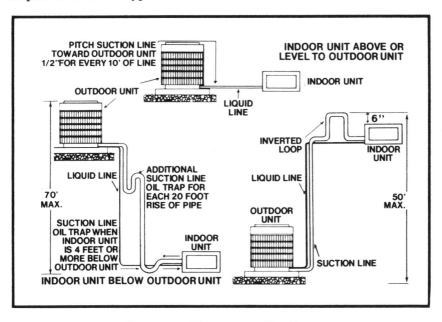

Figure 6.2   Oil return considerations.

6.3 REFRIGERANT TUBING

Most manufacturers offer optional, precharged tubing kits that are compatible with their heat pumps, although they can also be purchased separately from a variety of sources. The tubing in the indoor and outdoor units is normally evacuated and dehydrated at the factory and is then shipped with a holding charge. The vapor and liquid connections are sealed, as are the precharged tubing kits. Tubing should remain sealed until all tubing has been installed and connections are to be made. The manner in which precharged tubing is connected to the heat pump will vary depending on the types of fittings used.

The use of incorrectly sized tubing can cause a system to function below its rated capacity. Proper tubing size is checked using a chart such as that shown in Figure 6.3, which is normally included in the installation manual of the unit being installed. Correct tubing size is checked by determining the unit tonnage and the length of run to be used.

REFRIGERANT TUBING SIZES

NOMINAL UNIT TONNAGE	To 21 ft.		22 - 39 ft.		39 - 50 ft.		50 - 75 ft.	
	SUCT.	LIQ.	SUCT.	LIQ.	SUCT.	LIQ.	SUCT.	LIQ.
1 - 1/2	5/8"	1/4"	5/8"	1/4"	5/8"	1/4"	3/4"	3/8"
2	5/8"	1/4"	3/4"	3/8"	3/4"	3/8"	3/4"	3/8"
2 - 1/2	5/8"	1/4"	3/4"	3/8"	3/4"	3/8"	3/4"	3/8"
3	3/4"	3/8"	7/8"	3/8"	7/8"	3/8"	7/8"	3/8"
3 - 1/2	3/4"	3/8"	7/8"	3/8"	7/8"	3/8"	7/8"	1/2"
4	3/4"	3/8"	7/8"	3/8"	7/8"	1/2"	1 - 1/8"	1/2"
5	7/8"	3/8"	1 - 1/8"	3/8"	1 - 1/8"	1/2"	1 - 1/8"	5/8"

NOTE 1: For tubing lengths in excess of 50 feet, add 3 oz. of oil for every additional 10 feet.
NOTE 2: REFRIGERANT CHARGE ADJUSTMENT FOR TUBING LENGTHS BEYOND 25 FEET:

A. 3/8" OD LIQUID LINE - add 0.6 oz. F22 for each foot beyond 25 feet.
B. 1/2" OD LIQUID LINE - add 0.6 oz. F22 for each foot up to 25 feet and 1.2 oz. per foot for each foot beyond 25 feet.
C. 5/8" OD LIQUID LINE - add 1.2 oz. F22 for each foot up to 25 feet and 1.8 oz. per foot for each foot beyond 25 feet.

Figure 6.3   Refrigerant tubing sizes.

The following guidelines can be observed when installing refrigerant tubing, although each manufacturer's procedures must be followed. This is because the type of tubing and the manner in which it is connected to the system varies from manufacturer to manufacturer.

1. Before proceeding with any piping, connect a supply line voltage to the unit to energize the compressor crankcase heater. This will cause any refrigerant liquid that may have migrated to the compressor to be boiled off while the refrigerant mains are being connected.
2. Indoor and outdoor units should be connected together only with tubing that is compatible with the heat pump being installed.
3. Permanent isolation hangers should be used on both the liquid and suction lines (Figure 6.4 shows several different methods).
4. The suction line should be insulated to prevent sweating.

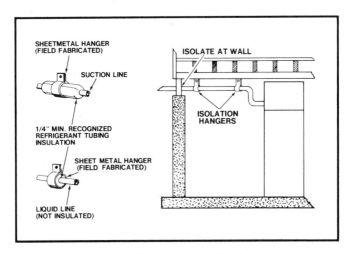

Figure 6.4   Permanent isolation hangers.

5. Caution should be exercised to ensure that the tubing does not touch walls and/or ceilings, to prevent transmission of noise to the structure when the unit is operating. When passing tubing through walls, seal the opening. Also, allow slack between the structure and the unit; this will aid in absorbing vibration.

6. Install lines so that they do not interfere with filter removal.

7. Run the lines with as few bends as possible, taking care not to damage fittings/couplings.

8. When making tubing connections, be sure to provide clearance at the unit for electrical connections.

9. Lines should be brought into the evaporator in such a manner as to avoid obstructing access to blower components.

10. Install a filter/drier in liquid line as dictated.

11. Any excess tubing should be located indoors and coiled in a horizontal position, with the suction line slanting toward the indoor unit.

12. Locations where the tubing will be exposed to mechanical damage should be avoided.

After observing these guidelines for hanging refrigerant tubing, follow the manufacturer's procedures for connecting the tubing to the heat pump unit(s).

Leak Testing

All refrigerant tubing should be exposed for visual inspection and leak tested before being covered or enclosed. All piping within the heat pump unit and cooling coil, as well as all interconnecting tubing, should be properly leak tested (see Chapter 4). If any leaks are discovered, pressure should be released and the leaks repaired, with the leak test being repeated as indicated.

Purging/Evacuating/Charging

Once the system is free of leaks, it will be necessary to determine whether it needs purging. Purging removes any contaminants that may have entered the system. This will be necessary if the tubing is not precharged or if the tubing or the indoor coil has been exposed to atmospheric conditions during installation, usually for longer than 5 minutes. Refer to Chapter 4 for standard purging procedures. After purging, leak testing should be repeated.

If the tubing or indoor coil are contaminated with moisture and/or liquid water during installation, both the tubing and the indoor coil must be evacuated. Evacuation serves to remove noncondensibles such as air, water, and inert gases, and assures a tight, dry system before charging. Refer to Chapter 4 for standard evacuation procedures. The system is then charged (see Chapter 4).

6.4 CONDENSATE DRAIN CONNECTION

A properly designed condensate drain must be installed at the indoor unit to provide a means of carrying away any water that may accumulate during unit operation. The drain line must be trapped, and it must be protected from freezing temperatures. Also, the drain line should be insulated, especially where moisture drippage would be objectionable or would cause damage.

6.5 AUXILIARY HEAT AND BALANCE POINT SETTING

Almost all air-source heat pumps require auxiliary heat, usually in the form of electric resistance heating elements that are installed in the indoor unit. More than one element is often required to provide sufficient auxiliary heat. The auxiliary system is turned on in stages, usually by means of an outdoor thermostat, which makes contact when a drop in outdoor temperature occurs. In some installations, more than one outdoor thermostat is used; each stage of electric heat after the first stage is governed by its own thermostat. The thermostat(s) are set progressively lower for each stage. When more than one thermostat is installed outdoors, a heat relay is required to provide a means of bypassing the outdoor thermostats when the heat pump breaks down and auxiliary heat is needed. Outdoor thermostats are also sometimes built directly into the outdoor unit, which will not require any field installation or wiring.

Electric heat elements are usually quite easy to install; they are simply "slid" into the compartment allotted for them in the indoor unit. The unit's installation manual should be referred to for specifics regarding the routing of the wiring of these units, because this portion of the installation will vary.

The balance point of a heat pump is the lowest temperature at which it is able to heat the structure unaided by auxiliary electric heat. The balance point is dependent on the outdoor design temperature, the structure's heat loss at the design temperature, and the unit's capacity. The balance point should be determined prior to installation (see Chapter 3). Once the balance point is determined, it is used to set the outdoor temperature at which the auxiliary heat will be energized.

6.6 ELECTRICAL CONNECTIONS

All field wiring must be performed in accordance with local codes. The voltage to the unit must be within permissible limits of voltages indicated on the nameplate of the unit being installed. If an improper line voltage is noted, the local power company should be consulted, because operation of a unit on an improper voltage can negate the unit's warranty.

Field wiring usually consists of installing disconnect switches, power supply wiring, and control wiring, following each manufacturer's specific instructions. This is important, since each heat pump's internal wiring may be different. The manufacturer will normally provide wiring diagrams that should be referred to while performing any field wiring.

The following guidelines can be referred to:

1. Provide a separate disconnect for the outdoor unit, the indoor unit, and for each accessory heater circuit. Locate disconnects within sight and ensure that they are readily accessible.
2. Ensure that the unit is properly grounded.
3. All electrical connections should be checked for size, proper installation and tightness before start-up.
4. All control wiring should be run separately to the control box.

6.7 THERMOSTATS

Heat pump manufacturers usually offer more than one indoor thermostat that is compatible with their system, providing the installer and/or homeowner with a number of choices. Only those thermostats recommended by the manufacturer of the heat pump being installed should be used.

An indoor thermostat should always be mounted approximately 5 feet from the floor and located where it will be exposed to normal room air circulation. It should not be located on an outside wall, near a supply air grille, or where it will be affected by sunlight or drafts. Also, circulation to the thermostat should not be blocked by curtains, drapes, furniture, or partitions.

Because of the wide variety of indoor thermostats manufactured today, it would not be practical or helpful to detail each individual installation procedure in this text. This portion of an installation is best accomplished with the aid of wiring diagrams and specific instructions provided with the unit being installed.

6.8 DUCTWORK

Air supply and return can be handled in a number of different ways, depending on the installation. It should be understood, however, that the vast majority of problems encountered with air-source heat pumps are due to an improperly designed or installed duct system. Therefore, it is crucial that the duct system be properly designed and installed.

The following guidelines should be observed:

1. On any installation, asbestos cloth collars or other nonflammable material should be used for the return air and discharge connections to minimize transmission of vibration.
2. The supply air opening on the indoor unit should be enlarged to the proper duct size by the use of a transition.
3. All ducts should be suspended using flexible hangers and never fastened directly to the structure.
4. Insulation of ductwork is necessary where it travels through an unheated space during the heating season or an uncooled space during the cooling season. The use of a vapor barrier is recommended to prevent absorption of moisture from the surrounding air into the insulation.

When the ductwork system is in place and the unit installed, it may be necessary to adjust the motor of the indoor blower assembly to assure balanced air distribution throughout the area to be heated or cooled. Such an adjustment is performed based on the air requirements of the unit and static resistances of both the supply and the return air-duct systems. This information is usually supplied in the form of tables with the unit being installed and should be used as a guide. Checking and adjusting air flow is usually accomplished with a manometer (see Chapter 4 for a discussion on the use of manometers).

6.9 START-UP

Pre-Start-up Checklist

Before start-up, the following checklist should be consulted to ensure that the system has been properly installed.

1. Thermostat is matched to heat pump equipment.
2. Indoor and outdoor units are properly matched.
3. Auxiliary heat is properly sized for structure, and proper outdoor thermostats (where applicable) have been installed.
4. All electrical connections are correct and tight.
5. Indoor blower is turning freely and its speed is correct.
6. Condensate drain pan is properly positioned and drain line is trapped, of the right size, and is clear.
7. If unit is installed in an unconditioned space, an auxiliary drain pan has been installed.
8. All supply and return air-duct connections are correct and tight.
9. All refrigerant tubing is correctly pitched, connected, and supported.
10. System has proper refrigerant charge and has been checked for leaks.
11. Outdoor fan is free-turning and outdoor thermostat(s) have been set to the proper balance point.
12. All ductwork has been properly sized and balanced, and all joints are tight.
13. All ductwork in unconditioned space is insulated.
14. All ductwork is properly supported.

Start-up Checklist

After performing all of the above checks, the system is ready for start-up. Because of the different types of controls in use today, it will again be necessary to refer to each manufacturer's specific start-up procedures; but the following guidelines can be observed:

1. Test the limit switch (let the unit come up to temperature with the blower off).
2. With the indoor blower on, make a system balance test.
3. Check amperage on blower motor to ensure that it does not exceed nameplate rating.
4. Turn on crankcase heater 12 hours before starting unit.

5. Start heat pump in proper mode (heating or cooling) and allow it to stabilize before check out.

6. Observe the operation of the compressor by listening for any unusual sounds.

7. Check voltage and amperage of compressor.

8. Check refrigerant tubing for vibration.

9. Check the refrigerant charge under the following conditions:
   a. Outdoor entering air—dry bulb
   b. Return entering air over coil—dry bulb
   c. Return air on indoor coil—wet bulb
   d. Low side (suction pressure)
   e. High side (head pressure)
   f. Suction-line temperature at outdoor unit at compressor before accumulator
   g. Hot gas line (on heating) leaving outdoor unit and indoor unit for heat sink (should be no more than a 5° loss). Refer to the manufacturer's specification sheets while noting the above results. Make adjustments as indicated.

10. Check operation of the thermostat.

11. Place unit into a manual defrost cycle and test for proper initiation and termination.

12. Adjust supply dampers to balance system and air flow.

If problems occur during start-up, refer to the information in Chapters 8 and 9 for troubleshooting and component-testing procedures.

Seven
Troubleshooting: Water-source Heat Pumps

This chapter provides an easy-to-use troubleshooting guide that will enable service personnel to quickly determine the problems with a water-source heat pump. After the cause for a malfunction has been determined, refer to Chapter 9 for detailed information on the repair of individual components within an air- or water-source heat pump system.

PROBLEM

Unit does not run.

Possible cause #1.   Blown fuse.

Checks and/or corrections.   Replace fuse or reset circuit breaker. Also check for correct size fuse.

Possible cause #2.   Broken or loose wires.

Checks and/or corrections.   Replace or tighten wires.

Possible cause #3.   Voltage supply low.

Checks and/or corrections.   If voltage is below minimum voltage specified on data plate of unit, contact local power company.

Possible cause #4.   Low voltage circuit.

Checks and/or corrections.   Check transformer for burnout or voltage less than specified for unit.

Possible cause #5. Thermostat.

Checks and/or corrections:

1. Set thermostat to "COOL" and lowest temperature setting. Unit should run.
2. Set thermostat to "HEAT" and highest temperature setting. Unit should run.
3. Set fan to "ON." Fan should run.
4. If unit does not run in all three of these settings, the thermostat may be wired incorrectly or may be at fault. Check thermostat operation and replace if necessary. Be sure the replacement is compatible with the heat pump unit.

PROBLEM

Unit operates, but does not cool properly.

Possible cause #1. Clogged air filter.

Checks and/or corrections. Check filter and replace if necessary.

Possible cause #2. Water flow through coil is restricted, stopped, or in insufficient supply.

Checks and/or corrections. Check water flow and increase gallons per minute if necessary.

Possible cause #3. Defective compressor or refrigerant leak.

Checks and/or corrections. If compressor runs but the evaporator coil does not cool down, this indicates either a defective compressor or loss of refrigerant charge. Perform checks on compressor and replace if necessary. Check refrigerant charge and recharge if necessary.

PROBLEM

Unit heats only.

Possible cause. Reversing valve.

Checks and/or corrections. The valve is de-energized due to improper wiring at the heat pump or at the thermostat. Correct wiring if necessary. If wiring is satisfactory, check valve operation and replace if necessary.

PROBLEM

Unit will not operate on call for heating.

Possible cause #1. Clogged or dirty air filter.

Checks and/or corrections. Check filter and clean or replace as indicated.

Possible cause #2. Thermostat improperly set.

Checks and/or corrections: Check thermostat setting to determine if it is set below room temperature. Adjust as indicated.

Possible cause #3. Defective thermostat.

Checks and/or corrections. Check thermostat operation and replace as indicated.

Possible cause #4.   Incorrect wiring.

Checks and/or corrections.   Check for broken, loose, or incorrect wiring. Make repairs as indicated.

Possible cause #5.   Blower motor defective.

Checks and/or corrections.   Check blower motor in one of the other switch positions. If fan does not operate, check for open overload. If motor is not overheated, replace it.

PROBLEM

Insufficient cooling or heating.

Possible cause #1.   Lack of sufficient water pressure, temperature and/or quantity of water. Possible scaling in water coil.

Checks and/or corrections.   Adjust flow according to specifications of the heat pump. Check coil for scaling and clean and descale as indicated.

Possible cause #2.   Unit undersized.

Checks and/or corrections.   Recalculate heat gains or losses for conditioned space. If excessive, rectify by adding insulation, shading, and so forth.

Possible cause #3.   Loss of conditioned air by leaks.

Checks and/or corrections.   Check for leaks in ductwork or introduction of ambient air through doors and windows. Repair as indicated.

Possible cause #4.   Improperly located thermostat.

Checks and/or corrections.   Check location of thermostat and relocate as indicated.

Possible cause #5.   Inadequate airflow.

Checks and/or corrections.   Check motor speed, duct sizing, and filter. Remove or add resistance accordingly. Change or clean filter.

Possible cause #6.   Loss of refrigerant charge.

Checks and/or corrections.   Check charge and adjust after adjusting cubic feet per minute (cfm) and gpm.

Possible cause #7.   Blower running backwards; indicates reversed capacitor leads in the blower motor.

Checks and/or corrections.   Check blower motor capacitor leads and reverse as indicated.

Possible cause #8.   Defective compressor.

Checks and/or corrections.   Check compressor. If discharge pressure is too low and suction pressure is too high, compressor may be defective and should be replaced.

Possible cause #9.   Defective reversing valve, creating bypass of refrigerant from discharge to suction side of compressor.

Checks and/or corrections.   Check operation of reversing valve and replace as indicated.

Possible cause #10.   Incorrect operating pressure.

Checks and/or corrections.   Refer to charts supplied with unit for correct operating pressure and adjust as indicated.

Possible cause #11.   Refrigerant system.

Checks and/or corrections.   Check strainer and capillary tubes for possible restrictions to refrigerant flow. The refrigerant system may be contaminated with moisture, noncondensibles, and/or particles. Dehydrate, evacuate, and recharge as indicated.

PROBLEM

Evaporator ices over.

Possible cause #1.   Clogged air filter.

Checks and/or corrections.   Check filter and replace as indicated.

Possible cause #2.   Motor set to wrong speed.

Checks and/or corrections.   Connect to higher speed tap.

Possible cause #3.   Evaporator blower motor tripping off on overload.

Checks and/or corrections.   Check for overheated evaporator blower motor and tripped overload. Replace motor as indicated.

Possible cause #4.   Low room temperature.

Checks and/or corrections.   Icing will occur if room temperature drops below $55°$F. De-ice as indicated.

Possible cause #5.   Water temperature is too low.

Checks and/or corrections.   This situation should rarely occur. De-ice as indicated.

PROBLEM

Blower motor runs but compressor does not.

Possible cause #1.   Thermostat.

Checks and/or corrections.   Check setting, calibration, and wiring of thermostat. Repair as indicated.

Possible cause #2.   Faulty wiring.

Checks and/or corrections.   Check for loose or broken wires at compressor, capacitor, or contactor.

Possible cause #3.   High or low pressure controls.

Checks and/or corrections.   The unit could be turned off by the high or low pressure cut-out control. Reset the thermostat to "OFF." After a few minutes, turn to "COOL." If the compressor runs, unit was off on high or low pressure. If the unit still fails to run, check for faulty pressure switch by jumping the high and low pressure controls individually. Replace as indicated.

Possible cause #4.   Defective lockout relay.

Checks and/or corrections.   Relay may be stuck open. Check by turning power off. If relay does not reset, it must be replaced.

Possible cause #5.   Defective capacitor.

Checks and/or corrections.   Check capacitor. Remove and rewire as indicated.

Possible cause #6.   Voltage supply low.

Checks and/or corrections.   If voltage is below minimum voltage specified on the data plate of the unit, contact local power company.

Possible cause #7.   Compressor overload open.

Checks and/or corrections.   If compressor is too hot to touch, the overload will not reset until the compressor cools down. If the compressor is cool and the overload does not reset, there may be a defective or open overload. If the overload is external, replace. Otherwise, replace compressor.

Possible cause #8.   Compressor motor grounded.

Checks and/or corrections.   Internal winding grounded to compressor shell. Replace compressor. If compressor burnout has occurred, install filter drier at suction line.

Possible cause #9.   Compressor windings open.

Checks and/or corrections.   Check continuity of compressor windings with an ohmmeter. If windings are open, replace compressor.

Possible cause #10.   Seized compressor.

Checks and/or corrections.   Try an auxiliary capacitor in parallel with the run capacitor momentarily. If the compressor starts but the problem recurs on starting, install an auxiliary start kit. If the compressor still does not start, replace.

PROBLEM

Unit short cycles.

Possible cause #1.   Thermostat.

Checks and/or corrections.   The differential is set too close in the thermostat. Readjust heat anticipator.

Possible cause #2.   Wiring and controls.

Checks and/or corrections.   Check for loose wiring or a defective control contactor. Repair wiring and replace contactor as indicated.

Possible cause #3.   Compressor overload.

Checks and/or corrections.   Check compressor overload and replace as indicated. If compressor runs too hot, it may be due to a deficient refrigerant charge. Check charge and recharge as indicated.

Possible cause #4.   Improperly located thermostat.

Checks and/or corrections.   Check location of thermostat and relocate as indicated.

PROBLEM

Unit off on high pressure cut-out control.

Possible cause #1. Discharge pressure too high on cooling cycle.

Checks and/or corrections. Check water flow and temperature and adjust as indicated. Also check water coil for scaling or plugging, and clean and descale as indicated.

Possible cause #2. Discharge pressure too high on heating cycle.

Checks and/or corrections. Check air flow and temperature. Check blower and filter. Check water coil for scaling or clogging.

Possible cause #3. Refrigerant charge.

Checks and/or corrections. The unit is overcharged with refrigerant. Bleed off some charge or evacuate, and recharge with specified amount of refrigerant.

Possible cause #4. Excessive air flow and insufficient water flow.

Checks and/or corrections. Check for correct cubic feet per minute (cfm). Check water flow for correct gpm. Make adjustments as indicated.

Possible cause #5. Defective high-pressure switch.

Checks and/or corrections. Switch may be stuck open, does not reset, or has defective calibration. Replace as indicated.

PROBLEM

Unit off on low pressure cut-out control.

Possible cause #1. Suction pressure too low on cooling cycle.

Checks and/or corrections. Check air flow and air temperature. Check blower, filter, and water coil. Make repairs and/or adjustments as indicated.

Possible cause #2. Suction pressure too low on heating cycle.

Checks and/or corrections. Check water flow and water temperature. Also check water coil. Adjust flow and temperature as indicated. Clean and descale water coil as indicated.

Possible cause #3. Refrigerant charge.

Checks and/or corrections. Low refrigerant charge. Locate leaks, repair, evacuate, and recharge with specified amount of refrigerant, being sure to check and adjust cfm and gpm first.

Possible cause #4. Defective low-pressure switch.

Checks and/or corrections. Switch may be stuck open, does not reset, or has defective calibration. Check and replace as indicated.

Eight
Troubleshooting: Air-source Heat Pumps

This chapter provides a troubleshooting guide for air-source heat pumps. The information has been compiled from many manufacturer's individual troubleshooting guides, and is designed to enable service personnel to quickly determine the problem with an air-source heat pump. Refer to Chapter 9 for detailed information pertaining to testing and servicing individual components.

PROBLEM

Unit does not run.

Possible cause #1. Blown fuse.

Checks and/or corrections. Replace fuse.

Possible cause #2. Power off or loose electrical connection.

Checks and/or corrections. Check for correct voltage at contactor in condensing unit.

Possible cause #3. Thermostat out of calibration, defective, or set too high.

Checks and/or corrections. Test and reset as indicated.

Possible cause #4. Defective transformer.

Checks and/or corrections. Check wiring and replace as indicated.

Possible cause #5. Compressor overload contacts open.

Checks and/or corrections. If external overload, replace. If internal overload, replace compressor.

Possible cause #6. High-pressure control open (if provided).

Checks and/or corrections. Reset.

PROBLEM

Unit operates, but does not cool properly and indicates low suction pressure.

Possible cause #1. Lack of refrigerant being returned to compressor.

Checks and/or corrections. Check refrigerant charge and leak test if leak is suspected.

Possible cause #2. Defective expansion valve.

Checks and/or corrections. Check and replace as indicated.

Possible cause #3. Clogged or dirty air filter.

Checks and/or corrections. Replace as indicated.

Possible cause #4. Fan running backwards; defective run capacitor; defective or improperly sized fan motor.

Checks and/or corrections. Check fan and all internal components. Repair or replace as indicated.

Possible cause #5. Restricted air flow.

Checks and/or corrections. Check air flow; adjust as indicated.

PROBLEM

Unit operates but does not cool sufficiently and indicates high suction pressure.

Possible cause #1. Overcharge of refrigerant.

Checks and/or corrections. Check charge and adjust as indicated.

Possible cause #2. Defective compressor valves.

Checks and/or corrections. Replace compressor.

Possible cause #3. Defective check or reversing valve; indicates leaking.

Checks and/or corrections. Check and replace as indicated.

Possible cause #4. Noncondensibles in system.

Checks and/or corrections. Purge or evacuate as indicated.

PROBLEM

Unit operates but does not cool sufficiently and indicates low liquid pressure.

Possible cause #1. Insufficient refrigerant charge or leak in system.

Checks and/or corrections. Check charge and adjust as indicated; leak test system and repair leaks as indicated.

Possible cause #2. Defective compressor valves.

Checks and/or corrections. Check valves and replace compressor as indicated.

Possible cause #3. Defective or improperly installed check or expansion valve.

Checks and/or corrections. Check installation and operation; repair or replace as indicated.

Possible cause #4. Restriction in liquid line coupling before gauge port.

Checks and/or corrections. Remove or replace as indicated.

Possible cause #5. Dirty indoor coil or filters.

Checks and/or corrections. Clean coil and/or replace filters as indicated.

PROBLEM

Unit runs but does not cool sufficiently and indicates high liquid pressure.

Possible cause #1. Overcharge of refrigerant.

Checks and/or corrections. Check charge and adjust as indicated.

Possible cause #2. Noncondensables in system.

Checks and/or corrections. Purge or evacuate as indicated.

Possible cause #3. Restriction in liquid line.

Checks and/or corrections. Replace defective fittings or components.

Possible cause #4. Outdoor fan (running backwards, defective run capacitor, improper motor rpms, loose blade).

Checks and/or corrections. Check motor for proper rpms; check run capacitor; check fan blades. Repair or replace as indicated.

Possible cause #5. Dirty outdoor coil.

Checks and/or corrections. Check outdoor coil and clean as indicated.

PROBLEM

Unit runs but does not heat sufficiently and indicates low suction pressure.

Possible cause #1. Refrigerant charge.

Checks and/or corrections. Check charge and adjust as indicated. Also leak test system and repair leaks as indicated.

Possible cause #2. Defective expansion valve.

Checks and/or corrections. Check and replace as indicated.

Possible cause #3. Restricted tubes at outdoor coil or couplings not completely open.

Checks and/or corrections. Check and replace components as indicated.

Possible cause #4. Defective outdoor fan/motor.

Checks and/or corrections. Check for proper rpms, check run capacitor, check to make sure fan is not running backwards. Repair and/or replace as indicated.

Possible cause #5.   Dirty outdoor coil.

Checks and/or corrections.   Check and clean as indicated.

PROBLEM

Unit runs but does not heat sufficiently and indicates high suction pressure.

Possible cause #1.   Overcharge of refrigerant.

Checks and/or corrections.   Check charge and adjust as indicated.

Possible cause #2.   Defective or leaking check valve or reversing valve.

Checks and/or corrections.   Check valve(s) and repair or replace as indicated.

Possible cause #3.   Defective compressor valves.

Checks and/or corrections.   Check valves and replace compressor as indicated.

Possible cause #4.   Noncondensibles in system.

Checks and/or corrections.   Purge or evacuate as indicated.

PROBLEM

Unit runs but does not heat sufficiently and indicates low liquid pressure.

Possible cause #1.   Insufficient refrigerant charge or leak in system.

Checks and/or corrections.   Check charge and adjust as indicated. Leak test system. Repair leaks as indicated.

Possible cause #2.   Defective compressor valves.

Checks and/or corrections.   Check valves and replace compressor as indicated.

Possible cause #3.   Restricted tubes or restriction in liquid line coupling.

Checks and/or corrections.   Check tubing and coupling. Repair or replace as indicated.

Possible cause #4.   Defective check or expansion valve.

Checks and/or corrections.   Check valve(s) and repair or replace as indicated.

PROBLEM

Unit runs but does not heat sufficiently and indicates high liquid pressure.

Possible cause #1.   Overcharge of refrigerant.

Checks and/or corrections.   Check charge and adjust as indicated.

Possible cause #2.   Noncondensibles in system.

Checks and/or corrections.   Purge or evacuate as indicated.

Possible cause #3.   Dirty indoor coil or filters.

Checks and/or corrections.   Check coil and filters. Clean and/or replace as indicated.

Possible cause #4. Blower motor.

Checks and/or corrections. Check for improper size and improper rpms. Be sure that motor is not running backwards.

Possible cause #5. Restricted air flow.

Checks and/or corrections. Check for covered or restricted return air grills and/or supply registers. Check for improperly sized ductwork. Check air flow and adjust as indicated.

PROBLEM

Unit runs but does not cool sufficiently and operating pressures are correct.

Possible cause #1.   Loose or faulty wiring/circuits or components.

Checks and/or corrections.   Check all wiring and repair or replace wiring or components as indicated.

Possible cause #2.   Unit improperly sized.

Checks and/or corrections.   May necessitate performing new load calculations and replacing original unit with properly sized unit.

Possible cause #3.   Air leakage.

Checks and/or corrections.   Check air flow and adjust or repair air system as indicated.

PROBLEM

Unit runs but does not heat sufficiently and operating pressures are correct.

Possible cause #1.   Outdoor thermostat improperly set.

Checks and/or corrections.   Check setting and adjust as indicated.

Possible cause #2.   Restricted air flow or leakage (may be caused by improperly sized ductwork or location of registers).

Checks and/or corrections.   Check air flow and ductwork system, making sure system is sealed and insulated in any nonheated spaces.

PROBLEM

Unit does not defrost completely in defrost cycle.

Possible cause #1.   Refrigerant charge.

Checks and/or corrections.   Check refrigerant charge and adjust as indicated.

Possible cause #2.   Compressor valves.

Checks and/or corrections.   Check compressor valves for leakage. Replace as indicated.

Possible cause #3.   Defective defrost circuitry and/or controls.

Checks and/or corrections.   Check pressure and temperature settings and repair or replace as indicated.

PROBLEM

Compressor cycles off when in defrost cycle.

Possible cause #1.   Defective expansion valve.

Checks and/or corrections.   Check expansion valve with gauges and thermometer. Repair or replace as indicated.

Possible cause #2.   Dirty indoor coil and/or filters.

Checks and/or corrections.   Check and clean or replace as indicated.

Possible cause #3.   Defective or leaking check or reversing valve.

Checks and/or corrections.   Check valves and repair or replace as indicated.

PROBLEM

Auxiliary heat does not come on when unit is in defrost cycle.

Possible cause.   Defective defrost circuitry and/or components.

Checks and/or corrections.   Check wiring and components (de-ice controls, transformers, wiring on low-voltage circuit, and so on). Repair or replace as indicated.

PROBLEM

Auxiliary heat stays on after termination of defrost cycle.

Possible cause #1.   Defective defrost circuitry and/or components.

Checks and/or corrections.   Check control relay, de-ice control, and contactors. Repair or replace as indicated.

Possible cause #2.   Thermostat.

Checks and/or corrections.   Check operation of thermostat while unit is in defrost cycle. Replace if indicated.

Possible cause #3.   Defective emergency heat relay.

Checks and/or corrections.   Check and replace as indicated.

PROBLEM

Defrost cycle will not terminate.

Possible cause #1.   Temperature bulb.

Checks and/or corrections.   Check bulb to determine if it is loose or uninsulated. Repair or replace as indicated.

Possible cause #2.   Defective defrost controls.

Checks and/or corrections.   Check and repair or replace as indicated.

Possible cause #3.   Refrigerant charge.

Checks and/or corrections.   Check charge and adjust as indicated.

Possible cause #4.   Defective expansion valve.

Checks and/or corrections.   Check valve and repair or replace as indicated.

Possible cause #5.   Dirty indoor coil or filters.

Checks and/or corrections.   Clean and/or replace as indicated.

PROBLEM

Unit does not defrost.

Possible cause #1.   Closed defrost or reversing valve relays.

Checks and/or corrections.   Check terminals and adjust as indicated.

Possible cause #2.   Obstructed pressure tube.

Checks and/or corrections.   Check for obstructions. Repair or replace as indicated.

PROBLEM

Units heats instead of cooling.

Possible cause #1.   Reversing valve.

Checks and/or corrections.   Check reversing valve relay and replace as indicated.

Possible cause #2.   Defective electric resistance elements.

Checks and/or corrections.   Check for amperage. If amperage is indicated, check indoor thermostat for shorts. Repair shorts or replace thermostat if defective.

Nine

Component Testing and Troubleshooting

In this chapter, we will discuss the individual components in a heat pump system and provide general troubleshooting information on each. It should be understood that this is a general discussion and that the manufacturer's information should always be consulted for specifics as they relate to the repair of individual products.

9.1 COMPRESSORS AND MOTORS

As has been stated, the great majority of compressors used in heat pumps are hermetically sealed, meaning they are not serviceable in the field. However, there are some checks that can be made to determine the condition of a compressor in order to make a decision as to whether it must be replaced.

Compressors depend on their valves to provide a seal between the high and low pressure sides of the system. If either set of valves becomes damaged, pumping capacity and system efficiency are reduced or eliminated. This condition can be checked using a gauge manifold set. If the suction pressure will not pull down or the discharge will not build up, and the system is properly charged, a "low or no pumper" is suspected. If, when the system is shut down, the gauge pressures equalize rapidly, the valves should be suspected. Low compressor running amperage in a loaded condition is another indication of this same problem.

Both a noisy compressor and a "low or no pumper" are generally the result of refrigerant overcharge. Floodback or slugging the compressor will allow liquid refrigerant to damage or destroy valves, valve seats, and oil passages in the compressor. A system should never be overcharged to the point where liquid refrigerant returns to the compressor.

Compressor motors will usually take a great deal of abuse before failing. The motor may fail "OPEN," where one or more winding turns part and shut down the machine. It may fail "GROUNDED," where the wires touch ground on the motor stator or crankcase; or it may fail due to an internal "turn-to-turn" short, where one or more winding turns short together. It is important to remember that many hermetic compressors have an internal line-break overload that is sensitive to current and winding temperature. It is suggested that this overload, if used, be checked before a compressor is condemned.

To perform such a check, an ohmmeter is used. Power should be removed, and the compressor should be cool if possible. Remove the compressor terminal box cover and all wires from the terminals. Set the ohmmeter to the R × 1 scale. Touch one meter probe to C (common terminal) and the other probe to R (run), then to S (start) terminals. If a low resistance is noted, the overload is not open and the problem is elsewhere. If infinite resistance is noted, the overload may be open. Move the meter probes to R and S. If low resistance is noted, the overload is open.

Assuming the overload is not open, proceed to check out the motor windings. Adjust the ohmmeter to the R × 100 scale. Check first for a grounded winding by grounding one probe on the compressor discharge line. Touch the other probe on C–S–R in turn. If at any of the three terminals the meter registers, that winding has shorted to ground and the compressor must be replaced. If no ground is indicated, set the ohmmeter to the R × 1 scale. Connect one probe to the common terminal. Touch the other probe first to the run terminal and then to the start terminal and note the resistance. If the meter reads zero or no resistance, the motor is shorted "turn-to-turn." If the meter does not read at all, one or more of the motor windings are "open." A readable low resistance should be found if the motor windings are good. Typical run-winding resistance is between 1/2 and 2 ohms. Single-phase motors usually have a common-to-start winding resistance 3 to 6 times greater than the common-to-run windings. Connecting the meter probes from run to start should give a reading of the total of the common-to-start and common-to-run readings; i.e.:

Common Run:	1 Ohm
Common Start:	4 Ohms
Run Start:	5 Ohms

Before operating the equipment, it is suggested that gauge manifold hoses be connected to the proper refrigerant lines. Air must be purged from the hoses to protect the refrigerant system from uncondensables and contaminants. It is also suggested that an ammeter be connected at the common terminal of the compressor to record starting amperage as power is applied. Read the Locked Rotor Amperage (LRA) from the rating plate. As the compressor starts, the ammeter will momentarily draw at least the LRA and then drop back to a normal level. Should a problem exist in the compressor, the ammeter will continue to read high until a breaker or an overload breaks the circuit. This is a help in pinpointing a problem if one exists.

Sealed-system Evacuation

When sealed system has been opened to repair a leak or replace a defective component, the system must be evacuated prior to recharging. If a deep vacuum pump and thermister vacuum gauge are available, a vacuum of 300 microns (μ) should be pulled on the system. When the 300 μ level has been reached, the vacuum pump should be valved off from the system and the system vacuum should not rise at a rate greater than 100 μ per hour. Should it rise at a faster rate, evacuation should be continued until the 100 μ rate can be achieved.

If excessive moisture is suspected of having entered the system, a drier must be installed in the common suction line of the compressor. The drier used should be one size larger than the unit capacity. If the unit has been previously equipped with a drier, the existing drier should be removed and replaced with a new one. The system should then be operated in the same mode as when the failure occurred (heating or cooling). After 2 hours of operation, the drier should be checked for excessive pressure drop. To ensure moisture removal from the system when evacuation is being performed in outdoor ambient temperatures below $40°$F, the unit must be heated. Cover the outdoor unit and use a portable heater to apply warmth to the unit. After evacuation, charge the system using the superheat charging method (covered in the next section in this chapter), close service valves, and remove service equipment.

Compressor Burnout Clean-up Procedure

When a compressor-motor burnout occurs, the system oil may have been contaminated with acid that was formed due to overheating of the motor. If this acid is allowed to remain in the system, it will attack the system components, causing subsequent compressor failures. Compressor burnout can often be detected by the burned odor that is present when the system is opened. However, a sample of oil must be removed from the compressor and tested. If this test indicates the presence of acid, the acid must be removed and the system thoroughly cleaned.

The filter drier method of system clean-up, when applied properly, is a fast and very efficient method. When using this method, it is advantageous from a cost standpoint to use the replaceable-core type of drier. Although initial cost is somewhat higher, the replacement cores are relatively inexpensive. The drier must be one size larger than the tonnage capacity of the unit being serviced. The following steps should be performed:

1. Remove the defective compressor.
2. Attach an R-22 cylinder to the suction line and purge the system in the reverse direction at full cylinder pressure to remove as much oil and solids as possible.
3. Install replacement compressor.
4. On those units equipped with a drier, the existing drier must be removed and replaced with a new one. Install a clean-up filter drier in the common suction line.
5. Attach suction and liquid lines to the compressor.
6. Leak test the system.
7. Evacuate the system.
8. Charge the system using the superheat charging method (see next section) or weigh in the charge.
9. The mode of operation should not be changed to prevent any contaminants from entering the system. Allow the unit to operate for a minimum of 2 hours to assure pick-up of all contaminants. Repeat procedure if contaminants are found to still be in the system.
10. Remove drier from unit prior to changing modes.
11. Charge the system (see next section).

System Charging

A chart similar to the one shown in Figure 9.1 is normally provided with all heat pump units and should be used to check the cooling performance of the unit. One of the most accurate means of adding or

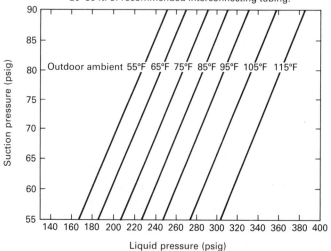

Figure 9.1   Normal operating pressure curve chart (cooling).

removing refrigerant from a capillary system is by the superheat method. This method allows the technician to accurately charge the system using the suction line pressure and temperature.

The superheat charging method is used only on cooling cycles. This method is accurate for all capillary-fed systems. Attach a thermometer or temperature tester sensing bulb to the suction line where it enters the condensing unit and insulate to ensure accurate readings. Attach a suction or compound gauge to the suction line port of the condensing unit. Start the unit and allow it to run until the system pressures stabilize. After pressures have stabilized, record the suction line pressure, suction line temperature, and the outdoor ambient. Enter on the charging chart that is provided with the unit (see sample in Figure 9.2) the actual suction line pressure and outdoor ambient, and compare the suction line temperature shown in the chart with the actual reading. If the intersection of temperature and pressure falls in the grey portion of the chart, the system must be checked for low air flow, refrigerant line restrictions, or other abnormal conditions. If the temperature is lower than that indicated on the chart, the system is overcharged and refrigerant must be purged. If the temperature is higher than that indicated on the chart, the system is undercharged and refrigerant must be added.

Heating performance charts such as those shown in Figure 9.3 are also provided with heat pump systems. Always refer to these charts for

Suction Line Temperature

Outdoor ambient (°F)	Suction pressure at outdoor section (PSIG)																								
	50	52	54	56	58	60	62	64	66	**68**	70	72	74	76	78	80	82	84	86	88	90	92	94	96	98
	Suction line temperature (+3°F.)																								
65	37	42	47	53	59	64	69	71	75	**79**	82	85													
70	32	37	43	48	53	59	64	69	71	**75**	79	82	85												
75		33	38	43	49	54	59	65	69	**72**	76	79	83	85											
80			32	38	43	48	54	59	64	**69**	72	75	79	81	85										
85			32	38	41	48	53	58	**64**	69	71	75	78	81	84										
90				**35**	**40**	**47**	**52**	**57**	62	65	68	72	75	78	81	85									
95						40	44	49	54	59	62	66	70	73	76	79	82	85							
100							44	48	53	58	63	65	69	72	75	78	81	85							
105							42	47	50	55	60	66	69	71	75	78	80	83	86						
110									45	50	55	60	65	68	70	73	76	79	82						
115											51	55	60	63	66	69	72	75	78	81	84				

NOTE: If suction line temperature falls in the gray areas of the chart, abnormal conditions exist. Check unit for low indoor unit airflow, refrigerant line restriction or other abnormal conditions, before adjusting charge. ALLOW SYSTEM PRESSURES TO STABILIZE BEFORE TAKING READINGS.

EXAMPLE: At 90°F outdoor temperature and 68 PSIG suction pressure, system will be correctly charged if the recorded suction line temperature is 57°F ±3°.

Figure 9.2   Suction line temperatures.

Typical Heating Performance Data Chart

Example:
At 50°F outdoor ambient and 70°F indoor temperature, typical suction and discharge pressures will be 53 PSIG and 296 PSIG respectively.

at 1600 CFM indoor airflow   at 1600 CFM indoor airflow

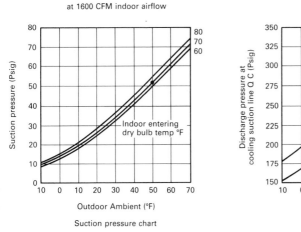

Suction pressure chart

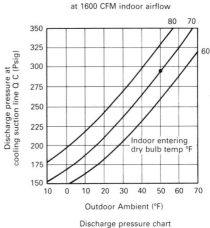

Discharge pressure chart

Figure 9.3   Heating performance chart.

the correct heating performance of the unit being serviced. Suction pressures during the heating cycle are obtained at the heating suction process port. To use heating performance charts, measure the outdoor and indoor ambients. Then refer to the charts to determine the suction and discharge pressures. The only accurate means of charging a heat pump at ambient temperatures below 65°F is by weighing in the refrigerant charge.

9.2 CONTROL BOX CIRCUITS AND COMPONENTS

Regardless of the type of heat pump, there will be a control box that contains most of the electrical components in the system. In a split system, the control box is located in the outdoor unit near the compressor compartment. Components such as contactors, capacitors, defrost controls, and start-assist devices are found in the control box of a heat pump. The control box also contains circuits and electrical components that operate the fan(s), motor(s), and/or blower assemblies.

The exact location of a control box and the positioning of circuits and components will vary depending on the manufacturer of the heat pump being installed. Wiring diagrams should always be referred to when troubleshooting the components in a heat pump's control box, but the following general discussion can also be referred to during diagnostic tests.

Capacitors

Capacitors are used in conjunction with compressor and/or fan motors to increase starting torque. In a control box, there may be a compressor motor run capacitor, a fan motor run capacitor, and possibly start capacitors as well. Since the operation of an electric motor can be improved or impaired by the condition of its capacitor(s), it is strongly suggested that the capacitor(s) be inspected at the same time as the motor it serves. Visually inspect each capacitor for bulged or leaking cases, and test each for "open" or "short."

To test a capacitor:

1. Discharge with 15,000 ohm, 2 watt resistor.
2. Remove all wires.
3. Set ohmmeter scale to R × 100.
4. Place probes across terminals.
5. The meter should move to a low resistance value, climb to a measurable resistance, and stop. If the needle does not move at all, the capacitor is open and must be replaced. If the needle moves to zero and stays there, the capacitor is shorted and must be replaced.
6. Check capacitance value. Use a capacitor tester if available. If not, use an ammeter and voltmeter. Place the voltmeter in parallel with

the charging circuit and the ammeter jaws around one leg of the charging circuit.

7. Energize the circuit only long enough to read meters. Use the following formula to compute capacitance:

$$(2650 \times \text{amps}) \div \text{volts}$$

Note: Capacitors should test within + or − 10% of case value.

Solid-state Motor Protector Modules

Some units incorporate a solid-state motor protector module in conjunction with the compressor. This module is actuated by internal sensors buried within the compressor motor windings to sense the winding temperature.

If a compressor does not operate and a solid-state module is incorporated, the module may be checked as follows:

1. Turn off electrical power to the unit.
2. Make sure the compressor has cooled sufficiently.
3. Place a jumper wire between the two terminals on the module, referring to wiring diagrams supplied with the module.
4. Apply electrical power and energize the control circuit by setting the appropriate controls to call for cooling. If the compressor contactor coil does not energize, the problem is elsewhere, possibly in the low-voltage electrical circuit.
5. Remove the jumper wire.
6. If the compressor operates, the sensors must be checked prior to concluding that the module itself is defective.

Internal Sensors in Solid-state Module

To check sensors:

1. Remove sensor leads from the module, marking each lead for replacement purposes.
2. Sensor circuits must be checked with an ohmmeter having a power source of 5 volts or less. Higher voltage will damage the sensors.
3. With an ohmmeter of proper range (0–200,000 ohms), measure the resistance through the sensor circuit. Refer to specifications provided with the module for acceptable resistance levels.
4. No meter reading indicates an open circuit. A meter reading of 0 (zero) ohms indicates a shorted or grounded circuit.

5. Should a sensor prove defective, the compressor must be replaced. If the sensors test good, the module itself is defective and is the only component that need be replaced.

Compressors with Start Assist

Some compressors utilize a solid-state variable resistor starting device in place of a relay and capacitor combination. The solid-state device performs in much the same manner as the relay and capacitor, in that it also supplies additional starting torque to the compressor motor. This device operates on its ability to change its internal resistance when a voltage is applied to it. If a compressor fails to start on the first attempt, the start assist will supply the compressor with another surge of voltage after a cool-down period of approximately 5 minutes.

To test the variable resistor:

1. Turn off electrical power to the unit.
2. Disconnect leads from the start device.
3. Check the device with an ohmmeter to determine if it is closed or open. No reading indicates an open. A meter reading indicates the circuit is complete.
4. If the device is found to be open, it is defective and must be replaced.

Start Kits

On some occasions, it may be necessary to use a start kit to obtain enough voltage to start the compressor. The start kit consists of an electrolytic capacitor, a potential relay, and mounting hardware. Other than using the commercial test instruments for capacitor testing, the start capacitor is tested in the same manner as outlined earlier. The start relay used in a start kit is checked in the same manner as outlined earlier under relays. Should a replacement kit be necessary, it should be matched to the existing equipment.

9.3 THERMOSTATIC EXPANSION VALVE

Thermostatic expansion valves are supplied for use with heat pumps by a number of manufacturers, although they all perform the same function in the same way. Figure 9.4 provides a troubleshooting chart that should be referred to when attempting to diagnose problems with a thermostatic expansion valve.

Low Suction Pressure, High Superheat

Probable Cause	Remedy
A. Expansion Valve Limiting Flow	
Inlet pressure too low from excessive vertical lift, undersize liquid line, or excessive low condensing temperature. Resulting pressure difference across valve too small.	Increase head pressure. If liquid line is too small, replace with proper size.
Gas in liquid line due to pressure drop in the line or insufficient refrigerant charge. If there is no sight glass in the liquid line, a characteristic whistling noise will be heard at the expansion valve.	Locate cause of liquid line flash gas and correct by use of any or all of the following methods: 1. Add charge. 2. Clean strainers, replace filter driers. 3. Check for proper line size. 4. Increase head pressure or decrease temperature to ensure solid liquid refrigerant at valve inlet.
Valve restricted by pressure drop through evaporator.	Change to an expansion valve having an external equalizer.
External equalizer line plugged, or external equalizer connector capped without providing a new valve cage or body with internal equalizer.	If external equalizer is plugged, repair or replace. Otherwise, replace with valve having correct equalizer.
Moisture, wax, oil, or dirt plugging valve orifice. Ice formation or wax at valve seat may be indicated by sudden rise in suction pressure after shutdown and system has warmed up.	Wax and oil indicate wrong type oil is being used. Purge and recharge system, using proper oil. Install a filter-drier to prevent moisture and dirt from plugging valve orifice.
Valve orifice too small.	Replace with proper valve.
Superheat adjustment too high.	Read about measuring and adjusting operating superheat.
Power assembly failure or partial loss of charge.	Replace power assembly (if possible) or replace valve.
Gas charged remote bulb of valve has lost control due to remote bulb tubing or power head being colder than the remote bulb.	Replace with a "W" cross ambient power assembly.
Filter screen clogged.	Clean all filter screens.
Wrong type oil.	Purge and recharge system and use proper oil.

Probable Cause	Remedy

B. Restriction in system other than expansion valve. (Usually, but not necessarily, indicated by frost or lower than normal temperatures at point of restriction.)

Probable Cause	Remedy
Strainers clogged or too small.	Remove and clean strainers. Check manufacturers catalog to make sure that correct strainer was installed. Add a filter-drier to system.
A solenoid valve not operating properly or is undersized.	If valve is undersized, check manufacturers catalog for proper size and conditions that would cause malfunction.
King valve at liquid receiver too small or not fully opened. Hand valve stem failure or valve too small or not fully opened. Discharge or suction service valve on compressor restricted or not fully opened.	Repair or replace faulty valve if it cannot be fully opened. Replace any undersized valve with one of correct size.
Plugged lines.	Clean, repair, or replace lines.
Liquid line too small.	Install proper size liquid line.
Suction line too small.	Install proper size suction line.
Wrong type of oil in system, blocking refrigerant flow.	Purge and recharge system and use proper oil.

Low Suction Pressure, Low Superheat

Probable Cause	Remedy
Poor distribution in evaporator, causing liquid to short circuit through favored passes and throttling valve before all passes receive sufficient refrigerant.	Clamp power assembly remote bulb to free draining suction line. Clean suction line thoroughly before clamping bulb in place. Install distributor. Balance evaporator load distribution.
Compressor oversize or running too fast due to wrong size pulley.	Reduce speed of compressor by installing proper size pulley, or provide compressor capacity control.
Uneven or inadequate evaporator loading due to poor air distribution or brine flow.	Balance evaporator load distribution by providing correct air or brine distribution.
Evaporator too small, often indicated by excessive ice formation.	Replace with proper size evaporator.
Excessive accumulation of oil in evaporator.	Alter suction piping to provide proper oil return or install oil separator, if required.

High Suction Pressure, High Superheat

Probable Cause	Remedy
Unbalanced system having an oversized evaporator, and undersized compressor and a high load on the evaporator. Load in excess of design conditions.	Balance system components for load requirements.
Compressor undersized.	Replace with proper size compressor.
Evaporator too large.	Replace with proper size evaporator.
Compressor discharge valves leaking.	Repair or replace valves.

High Suction Pressure, Low Superheat

Probable Cause	Remedy
Compressor undersized.	Replace with proper size compressor.
Valve superheat setting too low.	Adjust operating superheat.
Gas in liquid line with oversized expansion valve.	Replace with proper size expansion valve. Correct cause of flash gas.
Compressor discharge valves leaking.	Repair or replace discharge valves.
Pin and seat of expansion valve wire drawn, eroded, or held open by foreign material, resulting in liquid floodback.	Clean or replace damaged parts or replace valve. Install a filter-drier to remove foreign material from system.
Ruptured diaphragm or bellows in a constant pressure (automatic) expansion valve, resulting in liquid floodback.	Replace valve power assembly.
External equalizer line plugged, or external equalizer connection capped without providing a new valve cage or body with internal equalizer.	If external equalizer is plugged, repair or replace. Otherwise, replace with valve having correct equalizer.
Moisture freezing valve in open position.	Apply hot rags to valve to melt ice. Install a filter-drier to insure a moisture-free system.

Fluctuating Suction Pressure

Probable Cause	Remedy
Incorrect superheat adjustment.	Measure and adjust operating superheat.
Trapped suction line.	Install "P" trap to provide a free draining suction line.

Probable Cause	Remedy
Improper remote bulb location or installation.	Clamp remote bulb to free draining suction line. Clean suction line thoroughly before clamping bulb in place.
Floodback of liquid refrigerant caused by poorly designed liquid distribution device or uneven evaporator loading. Improperly mounted evaporator.	Replace faulty distributor. If evaporator loading is uneven, install proper load distribution devices to balance air velocity evenly over evaporator coils. Remount evaporator lines to provide proper angle.
External equalizer lines tapped at a common point although there is more than one expansion valve on same system.	Each valve must have its own separate equalizer line going directly to an appropriate location on evaporator outlet to ensure proper operational response of each individual valve.
Faulty condensing water regulator, causing change in pressure drop across valve.	Replace condensing water regulator.
Evaporative condenser cycling, causing radical change in pressure difference across expansion valve. Cycling of blowers or brine pumps.	Check spray nozzles, coil surface, control circuits, and thermostat overloads. Repair or replace any defective equipment. Clean clogged nozzles and coil surface.
Restricted external equalizer line.	Repair or replace with correct size.

Fluctuating Discharge Pressure

Probable Cause	Remedy
Faulty condensing water regulating valve.	Replace condensing water regulating valve.
Insufficient charge, usually accompanied by corresponding fluctuation in suction pressure.	Add charge to system.
Cycling of evaporative condenser.	Check spray nozzles, coil surface, control circuits, and thermostat overloads. Repair or replace any defective equipment. Clean clogged nozzles and coil surface.
Inadequate and fluctuating supply of cooling water to condenser.	Check water regulating valve and repair or replace if defective. Check water circuit for restrictions.
Cooling fan for condenser cycling.	Determine cause for cycling fan, and correct.

Probable Cause	Remedy
Fluctuating discharge pressure controls on low ambient air-cooled condenser.	Adjust, repair, or replace controls.

High Discharge Pressure	
Probable Cause	**Remedy**
Insufficient cooling water due to inadequate supply or faulty water valve.	Start pump and open water valves. Adjust, repair, or replace any defective equipment.
Condenser or liquid receiver too small.	Replace with correct size condenser or liquid receiver.
Cooling water above design temperature.	Increase supply of water by adjusting water valve, and replacing with a larger valve.
Air or noncondensable gases in condenser.	Purge and recharge system.
Overcharge of refrigerant.	Bleed to proper charge.
Condenser dirty.	Clean condenser.
Insufficient cooling air circulation over air-cooled condenser.	Properly locate condenser to freely dispel hot discharge air. Tighten or replace slipping belts or pulleys and be sure blower motor is of proper size.

Figure 9.4   Troubleshooting chart for thermostatic expansion valves (Courtesy Alco Controls Division, Emerson Electric Company).

Some problems with thermostatic expansion valves are related to the operating superheat and may require measuring and adjustment of the operating superheat. To measure superheat:

1. Determine suction pressure with an accurate gauge. On close-coupled installations, suction pressure may be read at the compressor suction connection.
2. From Figure 9.5, determine saturation temperature at observed suction pressure.
3. Measure the temperature of the suction gas at the remote bulb location as follows: Clean an area of the suction line at bulb location and tape thermocouple to cleaned area. Insulate thermocouple bead and read temperature on a potentiometer. If a potentiometer is not available, tape an accurate thermometer to the cleaned area

Bold Figures = Inches Mercury Vacuum   Light Figures = psig

°F	R-12	R-13	R-22	R-500	R-502	R-717 Ammonia
-100	27.0	7.5	25.0	—	23.3	27.4
-95	26.4	10.9	24.1	—	22.1	26.8
-90	25.7	14.2	23.0	—	20.7	26.1
-85	25.0	18.2	21.7	—	19.0	25.3
-80	24.1	22.2	20.2	—	17.1	24.3
-75	23.0	27.1	18.5	—	15.0	23.2
-70	21.8	32.0	16.6	—	12.6	21.9
-65	20.5	37.7	14.4	—	10.0	20.4
-60	19.0	43.5	12.0	—	7.0	18.6
-55	17.3	50.0	9.2	—	3.6	16.6
-50	15.4	57.0	6.2	—	0.0	14.3
-45	13.3	64.6	2.7	—	2.1	11.7
-40	11.0	72.7	0.5	7.9	4.3	8.7
-35	8.4	81.5	2.6	4.8	6.7	5.4
-30	5.5	91.0	4.9	1.4	9.4	1.6
-28	4.3	94.9	5.9	0.0	10.6	0.0
-26	3.0	98.9	6.9	0.7	11.7	0.8

°F	R-12	R-13	R-22	R-500	R-502	R-717 Ammonia
16	18.4	211.9	38.7	24.2	47.8	29.4
18	19.7	218.8	40.9	25.7	50.1	31.4
20	21.0	225.7	43.0	27.3	52.5	33.5
22	22.4	233.0	45.3	29.0	55.0	35.7
24	23.9	240.3	47.6	30.7	57.5	37.9
26	25.4	247.8	49.9	32.5	60.1	40.2
28	26.9	255.5	52.4	34.3	62.8	42.6
30	28.5	263.2	54.9	36.1	65.4	45.0
32	30.1	271.3	57.5	38.0	68.3	47.6
34	31.7	279.5	60.1	40.0	71.2	50.2
36	33.4	287.8	62.8	42.0	74.1	52.9
38	35.2	296.3	65.6	44.1	77.2	55.7
40	37.0	304.9	68.5	46.2	80.2	58.6
45	41.7	327.5	76.0	51.9	88.3	66.3
50	46.7	351.2	84.0	57.8	96.9	74.5
55	52.0	376.1	92.6	64.2	106.0	83.4
60	57.7	402.3	101.6	71.0	115.6	92.9

Temp						
-24	1.6	103.0	7.9	1.5	13.0	1.7
-22	0.3	107.3	9.0	2.3	14.2	2.6
-20	0.6	111.7	10.1	3.1	15.5	3.6
-18	1.3	116.2	11.3	4.0	16.9	4.6
-16	2.1	120.8	12.5	4.9	18.3	5.6
-14	2.8	125.7	13.8	5.8	19.7	6.7
-12	3.7	130.5	15.1	6.8	21.3	7.9
-10	4.5	135.4	16.5	7.8	22.8	9.0
-8	5.4	140.5	17.9	8.8	24.4	10.3
-6	6.3	145.7	19.3	9.9	26.0	11.6
-4	7.2	151.1	20.8	11.0	27.7	12.9
-2	8.2	156.5	22.4	12.1	29.5	14.3
0	9.1	162.1	24.0	13.3	31.2	15.7
2	10.2	167.9	25.6	14.5	33.1	17.2
4	11.2	173.7	27.3	15.7	35.0	18.8
6	12.3	179.8	29.1	17.0	37.0	20.4
8	13.5	185.9	30.9	18.4	39.1	22.1
10	14.6	192.1	32.8	19.8	41.1	23.8
12	15.8	198.6	34.7	21.2	43.3	25.6
14	17.1	205.2	36.7	22.7	45.5	27.5

Temp						
65	63.8	429.8	111.2	78.2	125.8	103.1
70	70.2	458.7	121.4	85.8	136.6	114.1
75	77.0	489.0	132.2	93.9	148.0	125.8
80	84.2	520.8	143.6	102.5	159.9	138.3
85	91.8	—	155.7	111.5	172.5	151.7
90	99.8	—	168.4	121.2	185.8	165.9
95	108.3	—	181.8	131.2	199.7	181.1
100	117.2	—	195.9	141.9	214.4	197.2
105	126.6	—	210.8	153.1	229.7	214.2
110	136.4	—	226.4	164.9	245.8	232.3
115	146.8	—	242.7	177.3	262.6	251.5
120	157.7	—	259.9	190.3	280.3	271.7
125	169.1	—	277.9	203.9	298.7	293.1
130	181.0	—	296.8	218.2	318.0	315.0
135	193.5	—	316.6	233.2	338.1	335.0
140	206.6	—	337.3	248.8	359.1	365.0
145	220.6	—	358.9	265.2	381.1	390.0
150	234.6	—	381.5	282.3	403.9	420.0
155	249.9	—	405.2	300.1	427.8	450.0
160	265.12	—	429.8	318.7	452.6	490.0

Figure 9.5 Temperature/pressure chart (Courtesy Alco Controls Division, Emerson Electric Company).

of the suction line. Insulate the thermometer bulb from ambient temperatures and observe the suction gas temperature.

4. Subtract the saturation temperature determined in Step 2 from the temperature measured in Step 3. The difference is the super-heat of the suction gas.

To adjust the superheat setting of an internally adjustable valve:

1. Pump the unit down and disconnect the outlet line from the valve.
2. Rotate the internal adjusting nut to increase or decrease superheat. Turning the nut clockwise will compress the valve spring and de-crease refrigerant flow through the valve, thus increasing superheat. Decompressing the spring by rotating the nut counterclockwise will increase flow through the valve and lower superheat.

If the valve is externally adjustable, it will not be necessary to re-move any line and thus, pump the unit down. To adjust: Remove seal cap from the bottom of the valve to expose the adjusting stem. Rotat-ing the stem clockwise decreases refrigerant flow through the valve and increases superheat. Rotating the stem counterclockwise increases flow through the valve and lowers superheat.

9.4 REVERSING VALVE

The reversing valve is an electrically controlled valve that is used to con-trol the flow of refrigerant through the system. The valve functions on a pressure differential and is therefore reliant on the system pressures to function properly. The reversing valve is placed in the system so that it is connected to the compressor discharge and suction lines.

If the reversing valve is suspected of being defective or inoperative, a number of checks, both on the valve and the system, should be com-pleted before replacing the valve. Check the system operating pressures prior to performing any checks on the reversing valve itself. If operating pressures are up to specifications, then the valve should be suspected.

Check the physical condition of the valve body for dents, deep scratches, or cracks. A damaged body can prevent the slide assembly from moving. The solenoid coil and needle valve are checked by ener-gizing the electrical system. To check the coil, first disconnect the con-denser fan motor and remove the fan blade. Next, with the solenoid coil energized, remove the coil lock nut and slide the coil partially off the coil stem. A resistance of the magnetic force should be felt if the coil is operative. Moving the coil further off the stem should result in a clicking of the plunger needle, indicating that the needle is responding to the solenoid coil's magnetic field. When returning the coil to its normal position on the stem, another clicking noise indicates that the plunger responded to the energized coil. If these conditions have not been satisfied, other components of the electrical system should be checked for possible trouble.

After all of the previous inspections and checks have been made and determined to be correct, perform the touch test on the reversing valve according to the chart in Figure 9.6. This test is performed by simply feeling the temperature relationships of the six tubes on the valve and comparing the temperature differences. Refer to the chart after the comparative temperatures have been determined for the possible cause and suggested corrective action to be taken.

TOUCH TEST CHART

VALVE OPERATING CONDITION	DISCHARGE TUBE from Compressor	SUCTION TUBE to Compressor	Tube to INSIDE COIL	Tube to OUTSIDE COIL	LEFT Pilot Capillary Tube	RIGHT Pilot Capillary Tube	NOTES: *Temperature of Valve Body. **Warmer than Valve Body.	
	1	2	3	4	5	6	Possible Causes	Corrections
NORMAL OPERATION OF VALVE								
Normal COOLING	Hot	Cool	Cool, as (2)	Hot, as (1)	*TVB	*TVB		
Normal HEATING	Hot	Cool	Hot, as (1)	Cool, as (2)	*TVB	*TVB		
MALFUNCTION OF VALVE								
Valve will not shift from cool to heat	Check electrical circuit and coil						No voltage to coil.	Repair electrical circuit.
							Defective coil.	Replace coil.
	Check refrigeration charge						Low charge.	Repair leak, recharge system.
							Pressure differential too high.	Recheck system.
	Hot	Cool	Cool, as (2)	Hot, as (1)	*TVB	Hot	Pilot valve okay. Dirt in one bleeder hole.	Deenergize solenoid, raise head pressure, reenergize solenoid to break dirt loose. If unsuccessful, remove valve, wash out. Check on air before installing. If no movement, replace valve, add strainer to discharge tube, mount valve horizontally.
							Piston cup leak.	Stop unit. After pressures equalize, restart with solenoid energized. If valve shifts, reattempt with compressor running. If still no shift, replace valve.
	Hot	Cool	Cool, as (2)	Hot, as (1)	*TVB	*TVB	Clogged pilot tubes.	Raise head pressure, operate solenoid to free. If still no shift, replace valve.
	Hot	Cool	Cool, as (2)	Hot, as (1)	Hot	Hot	Both ports of pilot open. (Back seat port did not close.)	Raise head pressure, operate solenoid to free partially clogged port. If still no shift, replace valve.
	Warm	Cool	Cool, as (2)	Warm as (1)	*TVB	Warm	**Defective Compressor.**	
Start to shift but does not complete reversal	Hot	Warm	Warm	Hot	*TVB	Hot	Not enough pressure differential at start of stroke or not enough flow to maintain pressure differential.	Check unit for correct operating pressures and charge. Raise head pressure. If no shift, use valve with smaller ports.
							Body damage.	Replace Valve.

VALVE OPERATED SATISFACTORILY PRIOR TO COMPRESSOR MOTOR BURN OUT—caused by dirt and small greasy particles inside the valve. **To CORRECT**: Remove valve, thoroughly wash it out. Check on air before reinstalling, or replace valve. Add strainer and filter-dryer to discharge tube between valve and compressor.

Figure 9.6   Touch test chart (Courtesy Command-Air Corporation; Waco, Texas).

Figure 9.6 (continued)

VALVE OPERATING CONDITION	DISCHARGE TUBE from Compressor	SUCTION TUBE to Compressor	Tube to INSIDE COIL	Tube to OUTSIDE COIL	LEFT Pilot Capillary Tube	RIGHT Pilot Capillary Tube	Possible Causes	Corrections
	1	2	3	4	5	6		
Start to shift but **does not complete reversal**	Hot	Warm	Warm	Hot	Hot	Hot	Both ports of Pilot open.	Raise head pressure, operate solenoid. If no shift, replace valve.
	Hot	Hot	Hot	Hot	*TVB	Hot	Body damage.	Replace valve.
							Valve hung up at mid-stroke. Pumping volume of compressor not sufficient to maintain reversal.	Raise head pressure, operate solenoid. If no shift, use valve with smaller ports.
	Hot	Hot	Hot	Hot	Hot	Hot	Both ports of Pilot open.	Raise head pressure, operate solenoid. If no shift, replace valve.
Apparent **leak** in heating	Hot	Cool	Hot, as (1)	Cool, as (2)	*TVB	**WVB	Piston needle on end of slide leaking.	Operate valve several times then recheck. If excessive leak, replace valve.
	Hot	Cool	Hot, as (1)	Cool, as (2)	**WVB	**WVB	Pilot needle and piston needle leaking.	Operate valve several times then recheck. If excessive leak, replace valve.
Will **not shift** from heat to cool	Hot	Cool	Hot, as (1)	Cool, as (2)	*TVB	*TVB	Pressure differential too high.	Stop unit. Will reverse during equalization period. Recheck system.
							Clogged Pilot tube.	Raise head pressure, operate solenoid to free dirt. If still no shift, replace valve.
	Hot	Cool	Hot, as (1)	Cool, as (2)	Hot	*TVB	Dirt in bleeder hole.	Raise head pressure, operate solenoid. Remove valve and wash out. Check on air before reinstalling, if no movement, replace valve. Add strainer to discharge tube. Mount valve horizontally.
	Hot	Cool	Hot, as (1)	Cool, as (2)	Hot	*TVB	Piston cup leak.	Stop unit, after pressures equalize, restart with solenoid de-energized. If valve shifts, reattempt with compressor running. If it still will not reverse while running, replace valve.
	Hot	Cool	Hot, as (1)	Cool, as (2)	Hot	Hot	Defective Pilot.	Replace Valve.
	Warm	Cool	Warm, as (1)	Cool, as (2)	Warm	*TVB	**Defective compressor.**	

NOTES:
*Temperature of Valve Body.
**Warmer than Valve Body.

To replace a reversing valve:

1. Remove the solenoid coil.

2. Heat the soldered joints to remove the valve. Exercise care to prevent the valve body from being heated above 250°F. Exceeding this temperature may introduce contaminants into the system. The valve body can be protected from excessive heat by wrapping it with wet cloths and keeping them wet during removal.

3. It is imperative when installing a new valve that the inside of all

tubes of the valve and the system be protected and kept clean of moisture and all other foreign matter.

4. The valve must be protected from excessive heat when installed.
5. Avoid rough handling of the replacement valve to prevent dents or bends from occurring on any portion of the valve.
6. Keep the axis of the valve body in a horizontal plane that may be rotated to any angle around its axis.
7. Make certain the limits of the valve capacity are not outside the specified refrigerant/tonnage of the system.
8. After the valve has been completely installed in the system and tested for leaks, return the solenoid coil to the pilot valve.
9. Recharge the unit with the weight of refrigerant specified by the manufacturer.
10. Determine that the unit is operating properly before attempting to operate the reversing valve. The valve will not operate properly on a partially charged system.
11. Cycle the valve at least a dozen or more times to check the proper operation of the system.

9.5 DEFROST CIRCUITRY, THERMOSTATS, AND AUXILIARY HEAT

As discussed in Chapter 3, defrost cycle initiation and termination can be accomplished in a number of different ways. Because methods and controls vary, it is recommended that the technician refer to specific information provided with the heat pump unit being installed if defrost components and/or circuitry are not operating properly.

The same holds true for thermostats and auxiliary electric heat elements and their associated circuitry. Both thermostats and auxiliary heat strips are offered as options by heat pump manufacturers, and they are responsible for assuring that these devices are compatible with their equipment, although such devices are usually supplied by another manufacturer. They are also responsible for providing the technician with associated specifications and wiring diagrams to aid in troubleshooting, should the devices fail or operate incorrectly. The manufacturer of the heat pump being installed is the best source for specifics regarding these devices.

Ten

Care
and
Maintenance

Periodic maintenance of a heat pump is the responsibility of the installer and the homeowner. Some maintenance procedures, of course, require the services of a technician and cannot be performed by the average homeowner, but homeowners should be advised as to their responsibilities at the time of installation so that they are aware of what needs to be done.

In this chapter, we will look at the standard care and maintenance procedures that need to be performed on a heat pump after it is installed. Those that are to be performed by a technician are covered first, followed by a list of items that the technician should explain to the homeowner regarding the operation, care, and maintenance of the heat pump system.

10.1 CHECKING AND CLEANING COILS

The outdoor coils of an air-source heat pump are exposed to dirt, and need to be cleaned on a regular basis. The indoor coil of both types of heat pumps should rarely need cleaning; an air filter is incorporated in the system to protect against dust and other foreign substances. The outdoor water coil of a water-source heat pump will be discussed in Section 10.2.

Coils should be checked on a monthly basis by the homeowner during both the heating and cooling seasons. They should be cleaned at least once at the beginning of each heating and cooling season to assure satisfactory operation, or more frequently as warranted.

The manner in which access is gained to a particular coil will depend on the design of the heat pump. However, it is imperative that a complete visual inspection be made, which will usually necessitate removing an access panel. Before this is done, all electrical power to the unit should be disconnected.

After power has been disconnected and the appropriate cover or access panel is removed, note the location of motor wires. If they are long enough to be moved out of the way, do so; otherwise, they may need to be disconnected. In the case of an outdoor coil, it may be cleaned from the inside outward with a garden hose. If this is not possible, the coil may be cleaned with a brush or a vacuum cleaner. During this procedure, care should be taken not to damage any of the coil's fins. Also, be sure to remove dirt from between coil rows.

If the coil is coated with oil or grease, it may be cleaned with a mild detergent or an approved coil-cleaning agent and then rinsed with clear water. Care should be taken not to get water near the compressor and/or control box. If it becomes necessary to clean the indoor coil of a heat pump, using a vacuum cleaner is usually most effective.

10.2 CLEANING AND DESCALING WATER COILS

Rust, scale, and slime on the surface of a water coil can interfere with the transfer of heat, reduce system capacity, cause higher head pressures, and increase the load on the system. Generally, these conditions will only occur in installations utilizing ground water drawn from a well, but a closed-loop system that utilizes water of high mineral content may also exhibit scaling in the water coil. The following procedure is recommended as a combined cleaning and descaling process that will quickly and economically restore the original operating efficiency of a water coil.

To perform this procedure, the following items will be required:

1. Oakite composition No. 22, available as a powder in drums.
2. Oakite composition No. 32, available as a liquid.
3. Clean water.
4. Acid proof pump and containers, or bottles with rubber hose.

Proceed as follows:

1. Drain and flush the water circuit of the coil. If scale and slime are found on the inner surfaces of the tube, a thorough cleaning is necessary before descaling can be accomplished.
2. To remove slime or mud, use Oakite composition No. 22, mixed 6 ounces per gallon of water. Warm this solution and circulate through the tubes until all slime and mud has been removed.
3. After cleaning, flush tubes thoroughly with clean water.
4. Prepare a 15% by volume solution for descaling by diluting Oakite compound No. 32 with water. This is accomplished by slowing adding a pint of the acid to 3 quarts of water. Since Oakite No. 32 is an acid, be sure that the acid is *slowly* added to the water. Do not put water into the acid, because this will cause spattering and excessive heat. Also, wear rubber gloves and wash the solution from the skin immediately if accidental contact occurs. Do not allow the solution to splash onto concrete.
5. Fill the tubes with the descaling solution from the bottom. Be sure to provide a vent at the top of the tubes for escaping gas.
6. Allow the Oakite No. 32 solution to soak in the tube coils for several hours, periodically pump-circulating it with an acid-proof pump. Alternatively, a bottle filled with the solution and attached to the coils by a hose can serve the same purpose. The bottle must be periodically raised and lowered to circulate the solution. The solution must contact the scale at every point for thorough descaling. Therefore, assure that no air pockets exist by regularly opening the vent to release gas. *Keep flames away from the vent gases.*
7. When descaling is complete, drain the solution and flush thoroughly with water.
8. Circulate a 2-ounce-per-gallon solution of Oakite No. 22 through the tubes to neutralize the acid. Drain this solution.
9. Flush the tubes thoroughly with fresh water.
10. Put the unit back in service and operate under normal load. Check the head pressure. If the pressure is normal, a thorough descaling has been achieved.

Water Coil Maintenance

To minimize scaling in a ground-water heat pump installation, the following precautions are recommended:

1. *Keep air out of the water.* Water should be checked to ensure that the well head is not allowing air to infiltrate the waterline. Lines

should always be airtight. Check by filling a container with water and adding water below the surface. Air will appear as fine bubbles or grey turbulence.

2. *Keep the system charged at all times.* It is recommended that a shut-off valve be placed in the discharge line to prevent loss of pressure between cycles.

3. *Keep a constant flow of water in the system.* Scaling from severe water conditions may be abated by keeping some water flowing through the system at all times. This will keep particles in suspension and thereby reduce the chances for scaling. This is accomplished by means of a valve that does not close completely, but allows a small amount of water to flow.

   If a unit is installed in known high mineral content water areas, it is best to establish with the owner a coil maintenance schedule so that the coil may be checked on a periodic basis. Cleaning may need to be done on a regular basis to assure maximum efficiency and prevention of permanent coil damage. Use the water coil cleaning method described in this section.

4. *Keep the gallons per minute (gpm) rate as high as possible.* Generally, the more water flowing through the coil, the less chance for scaling. A low gpm produces higher temperatures in the coil. However, do not exceed the gpm shown on the specification sheet for the unit installed.

10.3 COMPRESSOR LUBRICATION

All compressors require oil for lubrication. A sump area is provided in each compressor to store oil for lubrication, as shown in Figure 10.1. Oil circulates with refrigerant through the system, and it is important that the circulated oil returns to the compressor.

Liquid refrigerant dilutes oil and carries more oil through the system than does refrigerant gas. When liquid refrigerant is present in the compressor, it will dilute and remove large amounts of sump oil. Liquid refrigerant, therefore, should not be in the compressor. For this reason, installations of split systems must ensure proper return and retention of compressor oil. The correct use of traps and proper size tubing during installation will control both the oil and the liquid refrigerant. Poor piping practices could also result in quantities of oil collecting at the capillary tubes. The coils can become oil-logged, and suction pressure will slowly fluctuate from normal to low. With the compressor starved of oil, it can become noisy and may even fail.

All compressors and replacement compressors contain the proper

Compressor Lubrication

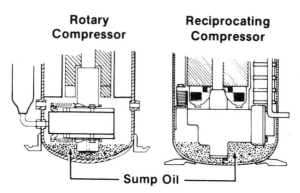

Figure 10.1   Compressor lubrication.

oil charge, and the addition of oil should not be necessary except when the oil is spilled. Installations requiring refrigerant lines of lengths of over 50 feet will require additional oil at a rate of .3 ounces for each 1 foot of line set addition. Where traps are used, their linear measurements are added to the line set length for the overall line set dimension.

If it becomes necessary to add oil to a system, the manufacturer of the compressor should be consulted. Different types of oils are used by different manufacturers, and one type of oil may not be supplemented by another type. The amount of oil to be added can be determined if a sight glass is available. If not, the manufacturer should be consulted for the proper amount of oil as well.

Thick, heavily discolored oil is an indication that contaminants are present in the system. All contaminated oil must be removed, the system cleaned, and new oil installed. The cause of contamination will need to be determined and corrected before installing new oil or the new oil will also become contaminated.

10.4 ELECTRICAL CONTROLS AND WIRING

All electrical connections in a heat pump system should be checked periodically to assure satisfactory operation. This procedure should be performed first with all power disconnected to the unit. An electrical check involves a visual inspection of all electrical connections, noting any smoky or burned connections. If problems of this nature are discovered, it will be necessary to clean all parts and stripped wire and reassemble them properly. Also, all screws on electrical connections should be tightened. To check all electrical controls, it will be necessary to reconnect electrical power to the unit and observe its operation through one complete operating cycle. If problems are apparent, utilize the troubleshooting information and component testing procedures outlined in Chapters 7, 8, and 9.

10.5 FAN MOTORS AND BLADES

Unless a fan motor actually fails, the only periodic maintenance that will be required is lubrication of the motor and cleaning of the fan blade. This procedure should be performed at least once a year.

Service will also be indicated if there is unusual noise during starting or operation. Power to the unit should first be disconnected, and the appropriate access panels removed to expose the fan motor and blades. For this procedure, it is best, if possible, to actually remove the assembly from the heat pump unit, taking care not to bend the fan blade. The fan can usually be removed without disconnecting the electrical leads to the motor because of the length of wire normally used. If, however, the leads are not long enough, it may be necessary to disconnect them.

To clean the motor and blade, use a soft brush or rag, taking care not to disturb the balance weights on the blade. It is also a good idea to check the fan set screw during this operation. To lubricate the motor, a good grade of SAE 20 nondetergent motor oil is used. Locate the oil holes and remove the plugs. A small amount of oil is then dropped into each hole, allowing time for the total quantity to be absorbed. Any excess oil is then wiped from the motor housing, and the plugs are replaced. To reassemble, simply replace the motor and blade, taking care that the electrical leads do not touch any refrigerant tubing. Reconnect electrical power and check the fan for proper operation. This procedure may be applied to any motor within a heat pump system.

10.6 AIR FILTERS

The homeowner should be advised that air filters need to be inspected on a monthly basis. Air flow is one of the most important aspects of efficient heat pump operation; dirty filters will cause serious problems by severely limiting air flow. Also, be sure to advise the homeowner that under no circumstances should the system ever be operated without an air filter. If air filters are properly maintained, the indoor coil of a heat pump system will rarely need cleaning.

10.7 ADVICE FOR THE HOMEOWNER

As part of an installation, the technician is responsible for advising the homeowner on the operation of the heat pump system, showing how to operate the controls, and providing a general list of tips on how to assure continued satisfactory operation.

The following list details the information that the technician should pass on to the homeowner to aid him or her in maintaining the heat pump system and avoiding unnecessary service calls:

1. Registers and return grills should not be blocked by furniture, draperies, or rugs in such a manner that air flow is restricted.

2. Although cutting off heat to unused rooms is frequently suggested as a conservation measure for conventional heating systems, this is not recommended for heat pump systems. All supply registers should be left open. A heat pump system is designed to heat or cool a specific space; compressor damage may occur if air circulation is reduced.

3. Nothing should interfere with ductwork in any area of the home, including basements and/or attics. Objects should not be placed on or near the ductwork in such a manner that it is disturbed in any way.

4. The drain pan should be periodically inspected and emptied during the summer months. Point out the location of condensate drain lines, and advise the homeowner that care should be taken that they are not disturbed in any way (bent, crimped, or crushed).

5. On air source units, the outdoor coils are exposed. Any sharp objects that are allowed to penetrate the coils can cause damage. Advise the homeowner to periodically check the outdoor unit to remove any leaves, paper, grass, or any other material that has accumulated and is blocking the unit.

6. The outdoor unit needs to be kept free of grass, weeds, bushes, and snow drifts.

7. The thermostat should be periodically checked for proper operation, and to ensure that the emergency heat switch has not been inadvertently turned on.

8. It is good practice to set the thermostat at the desired level of comfort and leave it there without trying to achieve comfort levels by manually changing the thermostat. Changing the thermostat setting to speed up cooling or heating will not be effective. A heat pump can take several hours to achieve the desired temperature setting; no amount of thermostat adjustment can speed this process.

9. Under no circumstances should the thermostat be turned off and then immediately turned on. This quick cycling may burn out controls in the starting circuits and in the compressor. Advise the homeowner that he or she should allow at least 5 minutes between all changes in thermostat settings.

10. Heat loss should be reduced as much as possible. Tight-fitting storm windows and doors should be installed, along with adequate insulation and caulking. Also, if a fireplace exists, advise the homeowner that as soon as a fire is out and the ashes have cooled, the flue should be closed to prevent heat loss.

11. When frost builds up and defrost is initiated in an air-source heat pump system, small amounts of cool air are sometimes felt at the supply registers; this is normal. Generally, the cool air is not released for a long enough time to cause discomfort. The homeowner should not adjust the thermostat in an attempt to alter this normal occurrence.

12. Advise the homeowner to listen for any changes in the normal operating sounds of the unit. Explain to the homeowner what short cycling is, and advise him or her that if this occurs, the equipment should be turned off until it can be serviced.

13. In the event of a power failure, the heat pump's switch *must* be turned off. After power is restored, the heat pump should then be restarted by setting the system switch to the desired operating mode.

14. Make certain that the homeowner has received and understands the warranty of the equipment installed.

A

Information Sources

Air Conditioning Contractors
of America (ACCA)
1228 17th Street, NW
Washington, DC 20036
(202) 296-7610

American Refrigeration Institute (ARI)
1501 Wilson Boulevard
Sixth Floor
Arlington, VA 22209

American Society of Heating,
Refrigeration and Air-
Conditioning Engineers
(ASHRAE)
1791 Tullie Circle, NE
Atlanta, GA 30329
(404) 636-8400

Charles Machine Works (Ditch
Witch Trenching Machines)
PO Box 66
Perry, OH 73077
(405) 336-4404

National Well Water Association
500 W Wilson Bridge Rd
Worthington, OH 43085
(614) 846-9355

Oklahoma State University
Engineering Technology
Extension
313 Crutchfield
Stillwater, OK 74078
(405) 624-5714

B
Water-source Heat Pump Manufacturers

COMPANY NAME	PRODUCT LINE
Addison Products Co PO Box 20434 Orlando, FL 32814 (305) 894–2891	WeatherKing
American Air Filter Allis-Chalmers Company PO Box 1100 Louisville, KY 40201 (502) 637–0325	EnerCon
Aqua-Matic Company PO Box 1046 Oklahoma City, OK 73101	Aqua-Matic
American Sun-Sol Samlin Enterprises 701 S Dixie Dr Vandalia, OH 45377 (513) 898–9733	Geo-Thermal Systems

COMPANY NAME	PRODUCT LINE
Bard Manufacturing Company PO Box 607 Bryan, OH 43506 (419) 636-1194	Bard
Borg-Warner Central Environ- mental Systems, Inc PO Box 1592 York, PA 17405 (717) 771-7890	Triton
Budco 6 Cadwell Rd Bloomfield, CT 06002	Solargy
California Heat Pump Company 2314 Michigan Ave Santa Monica, CA 90404 (213) 829-9275	California Heat Pump
Cantherm Heating Ltd 8089 Trans Canada Highway Ville St. Laurent, Quebec Canada H4S IS4 (514) 334-4879	Aquatherm; Terratherm
Carrier Air Conditioning Carrier Parkway Syracuse, NY 13221 (315) 432-6779	Weathermaker
Climate Control Snyder General Corp Residential Products Division 401 Randolph St Red Bud, IL 62278	ComfortMaker
Command-Aire Corporation PO Box 7916 Waco, TX 76714 (817) 840-3244	Command-Aire

COMPANY NAME	PRODUCT LINE
FHP Manufacturing Division of Leigh Products, Inc 601 NW 65th Court Fort Lauderdale, FL 33309 (305) 776-5471	Energy Miser
Friedrich Air Conditioning & Refrigeration Co 2007 Beechgrove Place Utica, NY 13501 (315) 724-7111	Geo-Thermal Heat Pump
GeoSystems, Inc 3623 N Park Dr Stillwater, OK 74074 (405) 372-6857	Closed-loop Systems
Heat Controller, Inc 1900 Wellworth Ave Jackson, MI 49203 (517) 787-2100	Comfort-Aire; Century
Heat Exchangers, Inc 8100 Monticello Ave Skokie, IL 60076 (312) 679-0300	Koldwave
Ice-Cap, Inc 48-25 36th St Long Island City, NY 11101	Ice-Cap
Lear Seigler, Inc Mammoth Division 13120-B County Road 6 Minneapolis, MN 55441 (612) 559-2711	Hydrobank; Sol-A-Terra
Marvair Company PO Box 400 Cordele, GA 31015 (912) 273-3636	Marvair; Crispair

COMPANY NAME	PRODUCT LINE
McQuay, Inc Air Conditioning Division 13600 Industrial Park Blvd Minneapolis, MN 55440	Seasonaire
National GeoThermal 1507 Buffalo Rd PO Box 703 Lawrenceburg, TN 38464 (615) 762-7106	Hydro-Solar
Northrup, Inc 302 Nichols Dr Hutchins, TX 75141 (214) 225-7351	En-Ex Systems
Phoenix Enviro-Tech 651 Vernon Way El Cajon, CA 92020 (714) 579-3883	Enviro-Temp
Prime Energy Systems Ltd 787 Alness St Downsview, Ontario Canada M3J 2H8 (416) 661-3303	Dantherm
The Singer Co. Climate Control Division 1300 Federal Blvd Carteret, NJ 07008 (201) 636-3300	Singer Heat Pump
SolarGen Corporation 1136 Wilmington Ave Dayton, OH 45420 (513) 258-0808	Solar/Geothermal Systems
Solar Oriented Environmental Systems, Inc 10639 Southwest 185th Terrace Miami, FL 33157 (305) 233-0711	Power Saver

COMPANY NAME	**PRODUCT LINE**
TempMaster Enterprises, Inc 1775 Central Florida Pkwy Orlando, FL 32809 (305) 851-9410	TempMaster
The Trane Company 3600 Pammel Creek Road La Crosse, WI 54601 (608) 787-3107	Trane
Trane Canada, Inc 401 Horner Avenue Toronto, Ontario Canada 14 M8W 2A5	Trane
Vanguard Energy Systems 9133 Chesapeake Dr San Diego, CA 92123 (714) 292-1433	**Vanguard**

C

Air-source Heat Pump Manufacturers

COMPANY NAME	PRODUCT LINE
Addison Products Company PO Box 63 Addison, MI 49220	**Weatherking**
Airtemp Air Conditioning 415 Wabash Ave Box 200 Effingham, IL 62401	**Airtemp**
Amana Refrigeration, Inc Amana, IA 52204	**Amana**
Arcoaire-Climate Control A unit of Snyder General Corp 302 Nichols Dr Hutchins, TX 75141	**Arcoaire**
Bard Manufacturing Co PO Box 607 Bryan, OH 43506 (419) 636-1194	**Bard**

COMPANY NAME	PRODUCT LINE
BDP Company 7310 W Morris St Indianapolis, IN 46206 (317) 243-0851	Bryant; Day & Night; Payne
Borg-Warner Central Environ- mental Systems, Inc PO Box 1592 York, PA 17405 (717) 771-7890	Luxaire; Moncrief; Fraser- Johnston; York
Bryant Air Conditioning	*See* **BDP Company**
Carrier Corporation 6304 Carrier Parkway PO Box 4800 Syracuse, NY 13221 (315) 433-4819	Weathermaker
Century By Heat Controller 1900 Wellworth Ave Jackson, MI 49203	Century
Climate Control A unit of Snyder General Corp 401 Randolph St Red Bud, IL 62278	AFCO; Comfortmaker; Climate Control
Coleman Company 250 North St Francis St Wichita, KS 67201	Deluxe Energy Saver
Daikin US Corp 910 Bern Court Suite 100 San Jose, CA 95112 (408) 280-6004	Daikin
Day & Night Air Conditioning	*See* **BDP Company**
Duo-Therm Corporation 509 S Poplar St LaGrange, IN 46761 (219) 463-2191	Conserver

COMPANY NAME	PRODUCT LINE
Fraser & Johnston	*See* Borg-Warner
Goettl Air Conditioning, Inc 2005 E Indian School Rd Phoenix, AZ 85064	Goettl
Goodman Manufacturing Co 1501 Seamist Houston, TX 77008	Janitrol
Heat Controller, Inc 1900 Wellworth Ave Jackson, MI 49203	Comfort-Aire
Heatwave International, Inc 10 N Elliott Aurora, MO 65605	Heatwave
Heil-Quaker Corporation 647 Thompson Lane Nashville, TN 37204	Heil
Home Division Lear Siegler, Inc 900 Brooks Ave Holland, MI 49423	Miller
Intertherm, Inc 3345 Morganford St. St. Louis, MO 63116	Weatherite
Lennox Industries, Inc Lennox Center 7920 Beltline Rd Dallas, TX 75240 (214) 980-6000	Lennox
Luxaire Heating & Air Conditioning	*See* Borg-Warner
Magic Chef Air Conditioning 851 W Third Ave Columbus, OH 43212 (614) 294-3547	Johnson; Magic Chef; Williams

COMPANY NAME	PRODUCT LINE
Marvair Company PO Box 400 Cordele, GA 31015 (912) 273-3636	Marvair; Crispair
Mitsubishi Electric Sales America, Inc 3030 E Victoria St Rancho Dominquez, CA 90221	Mitsubishi Electric
Moncrief Heating & Air Conditioning	See Borg-Warner
Patco, Inc 6955 Central Highway Pennsauken, NJ 08109	Patco
Payne Air Conditioning	See BDP
Rheem Manufacturing Co 5600 Old Greenwood Rd Forth Smith, AR 72906	Rheem; Ruud
Sanyo Electric Co 200 Riser Rd Little Ferry, NJ 07643	Sanyo
Sears, Roebuck & Co, Inc Sears Tower Chicago, IL 60684	Sears
Square D Company PO Box 766 Mesquite, TX 75149	Sun Dial
The Trane Company 3600 Pamel Creek Rd La Crosse, WI 54601 (608) 787-2000	Trane; GE Weathertron
Whirlpool Heating & Cooling Products 647 Thompson Lane Nashville, TN 37204	Whirlpool

COMPANY NAME	PRODUCT LINE
The Williamson Company 3500 Madison Rd Cincinnati, OH 45209	Williamson
York Heating & Air Conditioning	*See* Borg-Warner
Zoneaire Corporation PO Box 3517 Johnson City, TN 37601	Monterey

Index

A

Absolute zero, 2
Accumulator, 20, 56-57
Air filters, 191
Air-source heat pump:
  access to outside air, 25-27
  add-on split system, 25
  auxiliary heat, 148-49
  balance point setting, 148-49
  care/maintenance, 186-93
  charging, 148
  condensate drain connection,
    148
  cooling cycle, 26
  defrost cycle, 26
  ductwork, 150
  electrical connections, 149
  evacuation, 148
  heating cycle, 26

  leak testing, 147-48
  location/mounting, indoor
    unit, 144
  location/mounting, outdoor
    unit, 143-44
  oil return considerations, 145
  packaged, 25
  pre-start-up checklist, 151
  purging, 148
  refrigerant tubing, 145-48
  split-system, 25
  start-up checklist, 151-52
  supplemental heat, 26
  thermostats, 149-50
  troubleshooting, 159-65
Antifreeze, closed-loop systems,
    40, 136-37
Aquifer, 32
ASHRAE, 116-17
Auxiliary heat, 148-49, 185

Well pumps, 33
Wells:
  characteristics, 30-31, 119-20
  discharge, 30-33, 121
  existing, 30-31, 117-21
  flow requirements, 33, 117-18

  pumps, 33
  return piping, 33
  supply piping, 33
  supply, 30-33
  thermal interference, 33
  types, 117